CONTENTS OF VOLUME 1

SURFACE COATINGS—2

SURFACE COATINGS—2

Edited by

ALAN D. WILSON,
JOHN W. NICHOLSON

*Laboratory of the Government Chemist,
Department of Trade and Industry,
London, UK*

and

HAVARD J. PROSSER

*Warren Spring Laboratory,
Department of Trade and Industry,
Stevenage, UK*

ELSEVIER APPLIED SCIENCE
LONDON and NEW YORK

o3101101

CHEMISTRY

ELSEVIER APPLIED SCIENCE PUBLISHERS LTD
Crown House, Linton Road, Barking, Essex IG11 8JU, England

Sole Distributor in the USA and Canada
ELSEVIER SCIENCE PUBLISHING CO., INC.
52 Vanderbilt Avenue, New York, NY 10017, USA

WITH 27 TABLES AND 90 ILLUSTRATIONS

© 1988 ELSEVIER APPLIED SCIENCE PUBLISHERS LTD

British Library Cataloguing in Publication Data

Surface coatings.
2
1. Coatings
I. Wilson, Alan D. II. Nicholson, John W.
III. Prosser, Havard J.
667'.9

Library of Congress Cataloging-in-Publication Data

Surface coatings.

Includes bibliographies and indexes.
1. Coatings. I. Wilson, Alan D. II. Nicholson,
John W. III. Prosser, Havard J.
TP156.C57S87 1987   667'.9   87-8892

ISBN 1-85166-194-8 (v. 2)

Printed in Great Britain by Galliard (Printers) Ltd, Great Yarmouth

Preface

The science and technology of surface coatings continues to advance. Among the key areas are polymer chemistry, as new binders are developed to meet increasingly stringent environmental demands; testing and evaluation, as the need to understand the factors affecting coatings performance becomes ever more intense; and studies of that enduring problem, corrosion of metal substrates, from which coatings of ever-improving effectiveness are emerging. We have in this present volume of the series continued to cover aspects of these numerous developments. There are chapters on waterborne paint, a subject of increasing environmental importance, by J. W. Nicholson, and by H.-J. Streitberger and R. P. Osterloh; on a new and sophisticated test method, acoustic emission (R. D. Rawlings); and on anticorrosion coatings both organic (W. Funke) and inorganic (M. C. Andrade and A. Macias). Finally, that topic of immense practical importance to paint technology, pigmentation, is covered in a chapter by the late T. Entwistle.

All the authors have brought considerable experience in their chosen field of coatings technology to the preparation of their chapters, all of which are timely reviews of developing topics. We are grateful to each author for helping in the preparation of this volume, and for putting their experience at the disposal of the wide audience for whom this book is intended.

ALAN D. WILSON
JOHN W. NICHOLSON
HAVARD J. PROSSER

Contents

List of Contributors

M. C. ANDRADE

Instituto Eduardo Torroja de la Construcción y del Cemento, Apartado 19.002, 28080 Madrid, Spain

The late T. ENTWISTLE

Tioxide UK Ltd, Billingham, Cleveland TS23 1PS, UK

W. FUNKE

Forschungsinstitut für Pigmente und Lacke eV, Allmandring 37, D-7000 Stuttgart 80, Federal Republic of Germany

A. MACIAS

Instituto Eduardo Torroja de la Construcción y del Cemento, Apartado 19.002, 28080 Madrid, Spain

JOHN W. NICHOLSON

Laboratory of the Government Chemist, Department of Trade and Industry, Cornwall House, Waterloo Road, London SE1 8XY, UK

R. P. OSTERLOH

BASF AG, D-6700 Ludwigshafen, Federal Republic of Germany

REES D. RAWLINGS

Department of Materials, Royal School of Mines, Prince Consort Road, London SW7 2BP, UK

H.-J. STREITBERGER

BASF Lacke und Farben AG, Postfach 6123, D-4400 Münster, Federal Republic of Germany

CHAPTER 1

Waterborne Coatings

JOHN W. NICHOLSON

Laboratory of the Government Chemist, Department of Trade and Industry, London, UK

1. INTRODUCTION

Waterborne coatings are defined as 'coatings which are formulated to contain a substantial amount of water in the volatiles'.[1,2] At first sight, this definition does not appear to be particularly comprehensive or rigorous. However, the definition is based on very practical considerations, since the properties of a coating are determined largely by the nature of the solvent. In the case of waterborne coatings, the unique physical properties of water and its common occurrence define their use, method of application and exploitation. The fact that, as a class, such coatings include both true solutions and essentially insoluble suspensions, i.e. 'emulsion' or latex paints, is of secondary importance; in both cases, substantial quantities of water are used.

2. BACKGROUND TO THE DEVELOPMENT OF WATERBORNE COATINGS

Waterborne coatings are currently attracting much attention, and there have been many new developments in recent years.[3,4] The driving force for these developments is the need to restrict the release of organic solvents into the atmosphere, a fact which has also been responsible for the growth of the technologies of powder coatings and 'high-solids' coatings. Underlying the concern about solvent emissions are both environmental and economic considerations.

1

The main environmental concern has been in the United States, where there is now legislation governing atmospheric release of organic solvents, particularly olefins, aromatics and branched-chain ketones, which are photochemically active.[5] The original restrictions were embodied in the now-famous 'Rule 66', introduced by the Los Angeles County Authorities in July 1966. Los Angeles is particularly susceptible to environmental problems caused by release of organic solvents; because of the particular climatic features of this area, such emissions are not readily dispersed. Instead, they linger, and have led to severe photochemical smogs. In 1970, the US Government passed the Clean Air Act,[6] which laid down national standards for the control of hydrocarbon emissions, and within a decade all US states had individual regulations similar to those of 'Rule 66'.

Apart from the lack of environmentally damaging solvents, there are other attractions in the use of waterborne coatings. In particular, for the personnel working in paint application, there is less exposure to potentially harmful organic vapours, as well as less risk of fire.[7] In addition, waste disposal from paint spray lines using waterborne coatings is easier than disposal from corresponding lines using solvent-borne formulations.[7]

Economically, too, waterborne coatings are favoured. Not only is water the cheapest solvent available,[5] but its rivals, derived from crude oil, are subject to the erratic price of oil.[8]

As a result of all these pressures, and despite the widely recognised conservatism amongst paint users, waterborne coatings of various kinds are becoming increasingly important.[9] A recent questionnaire to the US coatings industry revealed that waterborne coatings were frequently judged to be the most important and pressing topic for research effort in the immediate future.[10] This interest is not confined to America. There continues to be a substantial and growing interest in waterborne coatings in

TABLE 1
Properties and Characteristics of Waterborne Coatings Systems (after Martens[1])

Property	Water-soluble	Emulsion (latex)
Appearance	Clear	Opaque
Particle size (μm)	0·001	0·1–1·0
Molecular weight	5 000–10 000	10^5–10^6
Viscosity	Depends on molar mass	Independent of molar mass
Film formation	Excellent	May need co-solvent
Viscosity control	Depends on molar mass	May need thickener
Gloss	Fairly high	Generally low

Europe, and this has been reflected in the estimate that they will achieve a market share of 25% by 1989.[11]

Broadly speaking, waterborne coatings fall between two extremes; these are (i) completely water-soluble and (ii) completely water-insoluble. In the latter case, the emulsion paints consist of a dispersion of solid resin in water; this is more correctly known as a latex.[1,2,11-14] Practical coating systems generally fall between these two extremes. For example, some resins are soluble only in mixed solvent–water systems (the so-called 'water-thinnable' or 'water-reducible' types), whilst other coatings may be formulated from a mixture of water-soluble and water-dispersed resins, and thus represent a sort of hybrid system.[12,15] In order to give an idea of the features of the different waterborne systems, the properties and characteristics of water-soluble and latex vehicles are compared in Table 1.

Having outlined the advantages, both environmental and economic, of waterborne coatings, it is also necessary to consider some of their drawbacks and the problems that still need to be solved. They include the following:

(i)   The final film has a tendency to remain water-sensitive.[5]

(ii)   More energy is needed to force-dry or stove waterborne coatings than solvent-based ones, possibly up to four times as much,[16] because of the high latent heat of evaporation of water.

(iii) Drying is heavily dependent on the humidity of the surrounding air. At low humidity, it may be explosively fast, and cause craters to appear in the final film.[17] Alternatively, at high humidity, for example in tropical countries such as Nigeria, where relative humidities of 78–85% are common,[18] it may be so slow that the coating undergoes severe sagging.[17]

(iv)   Flash rusting may occur on ferrous substrates.[19] Flash rusting is defined as the rapid corrosion of the substrate during drying of an aqueous coating, with the corrosion products (i.e. rust) appearing on the surface of the dried film. The phenomenon is not well understood, neither are the long-term effects on the stability of the coating known.

(v)   The quality of a waterborne coating is very dependent on surface cleanliness,[20] more so than for solvent-based systems. This is because of the high surface tension of water,[21] which causes it to wet poorly, especially greasy surfaces.

(vi)   Efflorescence may occur on certain substrates[22] and manifests itself as a growth of white crystals on wall surfaces as the paint dries out.[23] It is associated with the presence of water-soluble inorganic salts in the substrate, the salt being essentially Na_2SO_4 in gypsum plasters. Movement of these salts during drying of the plaster can cause severe disruption to the coating, and is a common source of paint failure in new buildings.[22]

(vii) Biocides are often required to preserve aqueous formulations.[1,2] This is because many of the components, such as the latex or binder, the fatty emulsifiers or defoamers, may be suitable nutrients for micro-organisms. Attack is enzymic, and may have a variety of possible consequences, such as loss of viscosity, gassing, smells or pH drift. The term 'biocide' covers both bactericides and fungicides, and includes a wide range of chemical species, from phenol derivatives to organotin compounds.[1]

(viii) Finally, emulsion paints, because of their nature, penetrate less well into porous surfaces than some other kinds of coating.[1] This assists what is called 'good holdout', but it can be a disadvantage. For example, if emulsion paints are used over old chalky paint surfaces, the lack of penetration may cause premature failure.[2]

3.   PRINCIPLES OF FILM FORMATION

In order for a waterborne formulation to function satisfactorily as a coating, it needs to be capable of forming coherent films, substantially free of voids, and to adhere well to the substrate.

The first of these requirements, the ability to form films showing good integrity, varies according to the nature of the polymer used. True solutions of polymer, generally with the aid of suitable surfactants, readily form continuous films, whereas latices may do so less readily, depending on the ease with which the polymer particles deform and coalesce. This whole question of film formation by latices is discussed in Section 6.1.

Adhesion, which is important in determining the performance of the final film, is generally regarded to depend on the ability of the initial coating to wet the substrate.[24,25] A large proportion of adhesion failures of coatings have been attributed to the inability of the coating to wet the substrate adequately, and it is appropriate to consider this problem in relation to the performance of waterborne coatings.

Classically, the ability of a liquid to wet a solid substrate has been considered in terms of surface tensions and contact angles.[26] When a drop of liquid is placed on a flat, solid surface, it generally remains as a drop

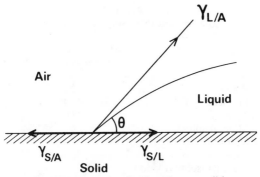

FIG. 1. Contact angle of a liquid on a solid.

having a distinct angle of contact with the solid surface, as illustrated in Fig. 1.

Taking the various surface forces as represented by the respective surface tensions acting in the direction of the surfaces, then equating the forces at the boundary of the solid, we obtain:

$$\gamma_{S/A} = \gamma_{S/L} + \gamma_{L/A} \cos \theta \qquad (1)$$

Finite contact angles arise when a liquid coheres more strongly to itself than it adheres to the surfaces of the solid. In principle, wetting can only occur when there is a contact angle of zero. In practice, however, spreading and wetting are found to occur for small finite values of θ, generally below 90°. This is because the theoretical model does not take account of the natural roughness of the solid surface, and it is this that aids wetting at non-zero values of θ.[27]

The actual value of the contact angle depends on whether the liquid is advancing or receding across the surface, this hysteresis effect being most apparent with impure surfaces. Contact angle also varies with surface contamination. For example, on a surface contaminated by the presence of grease, the contact angle for a water droplet increases, and wetting becomes less effective. Waterborne coatings are particularly sensitive to substrate quality in this way,[21] and this must be allowed for in considering their use.

Water has a very high surface tension, the highest of any solvent commonly used in coatings. This is illustrated in Table 2.

Mixtures of solvents do not usually exhibit additive surface tensions, particularly if their surface tensions are very different. Instead, what is usually observed is 'surface activity', which is the added liquid acting in the surface layers of the other to bring about a substantial reduction in the surface tension.

TABLE 2
Solvents and their Surface Tensions[27]

Solvent	Surface tension at 20°C $(mN\,m^{-1})$
Water	72·7
Ethylene glycol	48·4
Toluene	28·4
Mineral spirits	24
MiBK[a]	23·4

[a] Methyl isobutyl ketone.

If we consider again eqn (1), two facts are apparent. Firstly, the solid itself has a surface tension, $\gamma_{S/A}$, and secondly, there is a critical value of surface tension for liquids on a solid surface such that the contact angle is zero ($\cos\theta = 1$), and spontaneous wetting occurs.

To determine this value for a solid surface, contact angles for various liquid drops on the surface are measured, the cosines of the angles being plotted against the surface tensions of the various liquids. Extrapolation to $\cos\theta = 1$ (i.e. $\theta = 0$) gives γ_C. This method, utilizing the concept of critical surface tension (γ_C), provides a convenient method for assessing surface properties of various solids. Values of γ_C for different substrates are given in Table 3.

It is generally recognized that for good wetting (and hence good adhesion), the surface tensions of the liquid and solid should be similar to each other.[24] In order to apply waterborne coatings to the typical substrates listed in Table 3, the surface tension has to be reduced and this is

TABLE 3
Surface Tensions of Various Solids[25,27]

Substrate	γ_C at room temperature $(mN\,m^{-1})$
Zinc-phosphated steel	45–56
Aluminium	45
Iron-phosphated steel	43
PVC	39
Tinplate	35
Polyethylene	31
Untreated steel	29

usually done by using surfactants. Care needs to be taken with this, however, since too low a surface tension can lead to severe foaming.[14]

4. WATER AND ITS INTERACTION WITH POLYMERS

As a compound, water is remarkable. It is the only inorganic liquid to occur naturally on earth, and it is the only substance found in nature in all three physical states, solid, liquid and vapour.[28] It is the cheapest and most readily available solvent, and plays a vital role in the continuation of life on earth. Water circulates continuously in the environment by evaporation from the hydrosphere and subsequent precipitation from the atmosphere. This overall process is known as the hydrologic cycle. Reports estimate that the atmosphere contains about 6×10^{15} litres of water, and this is cycled 37 times a year to give an annual total precipitation figure of 224 × 10^{15} litres.[2,28] This precipitation does, of course, vary both geographically and seasonally.

In its physical properties, water is strikingly different from other solvents, not just in terms of its surface tension, as seen earlier, but also in its boiling point and latent heat of evaporation, its thermal conductivity and its density, all of which are very high for the molecular weight. These and other properties are listed in Table 4, together with those of other typical solvents used in the paint industry.

The high values of the various physical properties of water listed in Table 4 all point to the highly 'associated' nature of its constituent molecules.

TABLE 4
Physical Properties of Water and Other Solvents[29]

Property	Water	Mineral spirits (usually dodecane)	Acetone	Xylene
Molecular weight	18	170	58	106
Boiling point (°C)	100	214·5	56·5	144
Flash point (°C)	—	−12	−95	−25
Latent heat of evaporation at bp (kJ g^{-1})	2·259	0·481	0·565	0·39
Thermal conductivity × 10^3 (W m^{-1} °C^{-1})	5·8	1·49	1·8	1·59
Dipole moment (debyes)	1·84	0	2·88	0·4
Density (g cm^{-3})	1·0	0·752	0·787	0·86
Dielectric constant	78	1·83	21·3	2·37

Each water molecule consists of two hydrogen atoms covalently bonded to one oxygen atom. Both O–H bond lengths are 0·955 Å. The H–O–H angle is 104·5°,[30] which is slightly less than the ideal tetrahedral one of 109° because the lone pairs of electrons on the oxygen atom repel electrons more strongly than the bonding pairs between the oxygen and the hydrogen atoms.

In order to describe water, or any other liquid, in detail on the molecular level, one of two approaches may be adopted.[28] On the one hand, the problem can be tackled by considering the crystalline solid form of the compound, in which atoms and molecules occupy essentially fixed lattice sites, and the only movement allowed is oscillation about equilibrium positions. The liquid is then simply considered to be a highly perturbed solid, albeit much less ordered than before melting. The emphasis in this approach is placed on the positions occupied by molecules on a time-average basis.

On the other hand, it is possible to begin by considering a dilute gas whose molecules move about randomly and do not interact at all or influence the positions in space of each other. From this starting point a liquid may simply be regarded as a dense gas. Molecular motion is still important, though it is hindered by the relatively close presence of a substantial number of other molecules. In addition, there is considerable interaction between the molecules.

These two approaches are both useful, but at present, with the development of the techniques of X-ray and neutron diffraction, the emphasis is being placed on the first of them. Those features of liquids that resemble solids are stressed, and it is becoming acceptable to talk about the 'structure' of a liquid. This concept of the structure of a liquid at first sight appears odd. We generally use the term 'structure' for atoms and molecules arranged in a basically regular way, for example in a crystal. A liquid, by contrast, seems better characterized by randomness, there being substantial molecular motion, as shown for example in the Brownian motion of pollen grains suspended in water.

In order to obtain information on the structure of solids, the techniques of X-ray and neutron diffraction have been widely used. When these same techniques are applied to liquids, these are found to behave similarly to solids, scattering radiation in well-defined patterns of intensity. To process such data for liquids is more complex than for solids, because of the greater motion of the molecules, but it can be done, and it is clear from such experiments that liquids possess a degree of long-range order that can correctly be described as 'structure'. The existence of a detectable structure in liquid water is attributed to the presence of hydrogen bonds between adjacent water molecules. The difference between the electronegativity of oxygen and hydrogen leads to the polarity of the O–H bonds, and this in turn leads to the intermolecular force known as the hydrogen bond. The basic associated unit in water is the dimer, which can be represented thus:

$$\begin{array}{cc} H & H \\ \diagdown & \diagup \\ O\!-\!H\text{----}O \\ & \diagdown \\ & H \end{array}$$

Such units can themselves associate, and in the usual crystalline form of ice, where there is little movement of the molecules, this is the basis of the crystal structure. However, with four-coordinate oxygen atoms, the resulting structure is not a compact one, and as a result the structure of ice is sensitive to pressure. As well as the normal form, known as ice-I, there are eight other crystalline modifications, whose formation depends on the amount of external pressure.[28]

In liquid water, crystalline regions resembling ice-I form briefly in small areas of space; however, the kinetic energy of neighbouring molecules causes these 'quasi-crystalline' regions to dissolve quickly and then reform elsewhere. The essential difference, then, between ice-I and liquid water is that such crystalline regions are very much larger and longer-lived in the solid.[30]

The open structure of ice is responsible for two of the best-known anomalous properties of water, namely its increase in density on melting, and the achievement of maximum density 4°C above its freezing point. This behaviour results from the collapse of the relatively open hydrogen-bonded structure to a state where the molecules of liquid water are less regularly packed. There remains, though, considerable hydrogen bonding and short-range order, and this gradually breaks down as the temperature increases.[30]

4.1. Principles of Solubility/Dispersibility in Water

The general criterion for solubility is that 'like dissolves like',[31] i.e. polar solvents dissolve polar and ionic solutes, non-polar solvents dissolve non-polar solutes. In the case of water, this means that ionic compounds, such as sodium chloride, and polar compounds, such as ethanol, are soluble, but non-polar compounds, such as paraffin wax, are not.

In general, solubility depends on the relative magnitudes of three pairs of interactions, namely solvent–solvent, solute–solute and solvent–solute.[32]

For a substance to be soluble in a given liquid, the solvent–solute interactions must be greater than or equal to the interactions within each of the individual substances. If the internal forces of attraction within one or both exceeds that of the external interactions, the molecules of the different substances will remain close together, and will not mix with the dissimilar molecules.

When we deal with polymers, we consider the interactions of each individual segment of the macromolecule. A segment corresponds roughly to a monomer unit, and the same general principles governing solubility apply. Molecules made up of sufficiently polar segments are water-soluble; those made up of less polar or non-polar segments are not. To illustrate the kind of segments that confer water-solubility, three of the most widely used water-soluble polymers are shown in Fig. 2.

In practice, copolymers are often used in order to achieve the right balance of physical properties for a particular application, and highly polar monomers are included in order to confer either water-solubility or water-dispersibility. These two properties are related: for a polymer to be water-dispersible, its surface must be wettable by water. This is achieved by having a proportion of polar segments in the macromolecule, though less than required to confer complete water-solubility.

Neutralizing the polar groups where possible enhances the water-solubility, since it increases the hydrophilic nature of some of the polymer segments. For example, the copolymer of acrylic acid and methyl acrylate (11% methyl acrylate) is not naturally water-soluble, but it becomes so on neutralizing with sodium ions.[33] Some interesting anomalies are encountered when considering the water-solubility of certain polymers. Poly(vinyl alcohol), for example, which is prepared by hydrolysing poly(vinyl acetate), is not water-soluble when hydrolysis is complete.[34]

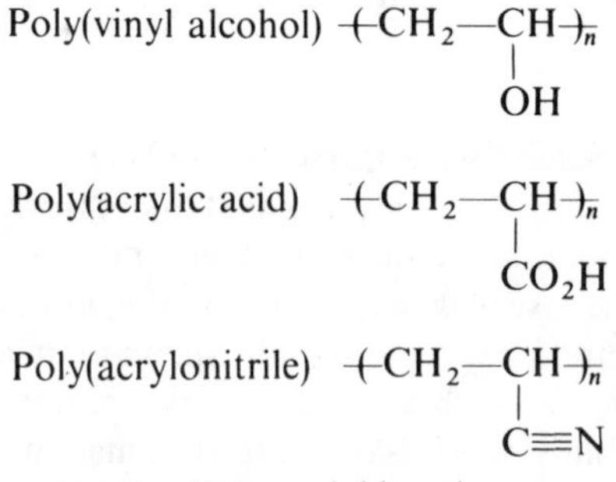

Poly(vinyl alcohol) $+CH_2-CH+_n$
                                              |
                                             OH

Poly(acrylic acid) $+CH_2-CH+_n$
                                          |
                                         CO_2H

Poly(acrylonitrile) $+CH_2-CH+_n$
                                          |
                                        $C\equiv N$

FIG. 2.   Water-soluble polymers.

However, it is very water-soluble if a few acetate groups are left unreacted. Ironically, the presence of a few relatively non-polar groups makes it more soluble. The reason is that such non-polar groups disrupt the regularity of the structure, thus enabling water molecules to enter and solvate the polar functional groups. Similarly, cellulose, which is a polymer of D(+)glucose,[35] a highly water-soluble compound, does not dissolve. Like poly(vinyl alcohol) it is simply too well ordered to permit water to enter the structure; it is insoluble for kinetic rather than thermodynamic reasons.

4.2. Hydrophobic Interactions in Polymer/Water Systems

So far, the discussion of solubility of polymers has been confined to a qualitative consideration of the attractive forces involved. However, all the polymers of practical use in coatings have hydrocarbon backbones, and in the absence of specific polar functional groups such structures would not be water-soluble. Thus, there is a 'hydrophobic' contribution to the overall solute–solvent interaction.[28] The origin of this hydrophobicity may be explained in terms of thermodynamic concepts.

Measuring enthalpy changes for the dissolution of hydrocarbons, such as alkanes, in water shows that ΔH is negative; energetically, water and alkanes attract one another. However, this does not make alkanes soluble in water to any appreciable extent, and this lack of solubility is not merely a kinetic effect. The free energy change, $\Delta G_{\text{Solution}}$, opposes the process, i.e. it is positive.

From

$$\Delta G_{\text{Solution}} = \Delta H_{\text{Solution}} - T\Delta S_{\text{Solution}}$$

it follows that the $T\Delta S_{\text{Solution}}$ term (and hence $\Delta S_{\text{Solution}}$) must be negative. This, in turn, leads to the conclusion that the entropy of the final state (i.e. the solution) must have decreased relative to the initial (i.e. two-phase) state, i.e. the proposed solution is more ordered than pure water. This result is attributed to the formation of 'cage' structures of hydrogen-bonded water molecules around the non-polar molecule in which the water has fewer degrees of freedom than in pure water itself.[28]

These hydrophobic interactions, arising on entropic grounds, are significant in solutions of all water-soluble polymers except poly(acrylic acid) and poly(acrylamide), where large $\Delta H_{\text{Solution}}$ terms swamp the effect.

In liquid–liquid phase diagrams, the hydrophobic interaction results in the existence of a lower critical solution temperature, and in the remarkable result that raising the temperature reduces the solubility (Fig. 3a). In general, the solution behaviour of water-soluble polymers represents a

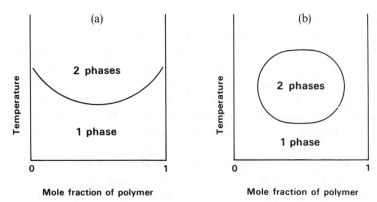

FIG. 3.   Phase diagrams of polymers in water.

balance between the hydrophilic and hydrophobic nature of the macromolecule. This often results in closed solubility loops (Fig. 3b), in which the lower-temperature behaviour and lower critical solution temperature arise due to hydrophobic effects, while, at higher temperatures, the solution behaviour becomes dominated by hydrophilic effects.

4.3 Emulsions and Latices
The coatings widely referred to as 'emulsion paints' are not in fact formulated from true emulsions at all, but are prepared from dispersions of insoluble polymer in water correctly called 'latices'.[23] The word 'emulsion' actually means a dispersion of one liquid in another, one of the liquids generally being water. A good example is milk, which consists of fat droplets dispersed in water.[36] In this case water is said to be the continuous phase, and fat the dispersed phase; either liquid may constitute either phase. Hence, emulsions can be either 'oil-in-water' or 'water-in-oil'. Such genuine emulsions are usually opaque, due to differences in the refractive index between the two liquids, and they generally require the presence of a dispersing agent to maintain their stability.

The correct name for a dispersion of solid polymer in water, 'latex', was originally applied to the milky white liquid that could be obtained from the natural rubber tree, *Hevea brasiliensis*.[37] Rubber latex consists of raw rubber, mainly *cis*-poly(isoprene), of molecular weight 300 000–500 000 in suspension in water. It has the approximate composition shown in Table 5.

The use of synthetic latex in paints began soon after the Second World War. The latex was produced by the polymerization of a true emulsion of monomer in water, and was originally used in the USA to produce a

TABLE 5
Approximate Composition of Natural Rubber Latex[37]

Water	60%	Sterols, fats, soaps	1%
Poly(isoprene)	35%	Inositol 1-methyl ether	1%
Proteins	2%	Inorganic salts	0·4%

synthetic rubber for tyres.[1] This synthetic rubber, known as Government Rubber Styrene, GRS, was needed because of the loss of supplies of raw natural rubber during the war; it was composed mainly of butadiene and styrene in the ratio 3:2. This particular latex, as it stood, was not suitable for use in paints, but reversing the ratios of the main components gave a harder copolymer, which was much more useful. In 1948, paints based on these latices were introduced on to the US market.

Most modern latex paints are of the vinyl type, being generally based on vinyl acetate, together with a variety of other monomers, such as ethyl acrylate, butyl maleate, ethylene,[23] acrylonitrile or styrene.[38] Copolymers are used rather than homopolymers in order to control the flexibility of the final film, and because they show better adhesion to surfaces and are able to tolerate greater pigment loading. The dispersed particles in either homopolymer or copolymer latices are negatively charged, and hence such latices are described as anionic dispersions.[23]

5. THE PREPARATION OF SYNTHETIC POLYMER LATICES

Since latex-based paints were first introduced on to the market in the late 1940s, they have been subjected to much development work. As a result, knowledge and understanding of polymerization techniques has grown, and this in turn has enabled manufacturers to prepare latices having properties tailored for specific end-uses. In this section, those principles underlying the preparation and stabilization of synthetic latices relevant to their use in surface coatings are described.

Synthetic latices are prepared by emulsion polymerization of appropriate monomers. The monomers, which are liquid, are dispersed in water with the aid of surfactants to form a true emulsion. The key requirement of such a system is stability. It is relatively easy to prepare an emulsion by vigorous agitation, but the mixture fairly soon separates out into its components on standing. Surfactants are used to prevent this happening.

Surfactants generally comprise long molecules having a hydrocarbon tail

which is hydrophobic, and an ionic or polar head, which is hydrophilic.
When present alone in water, with no monomer or other component, they
form micelles; a micelle is an aggregate of surfactant molecules, with their
hydrophobic tails oriented inwards towards each other, and their
hydrophilic heads outermost towards the water. When monomer is also
present, it tends to seek out the hydrophobic microphases inside the
micelles, causing them to swell up. The presence of the hydrophilic outer
shell causes the micelle to remain stable in water, partly by allowing the
overall particle to be hydrated, and partly by preventing too close an
approach or agglomeration between adjacent micelles. In the polymerization, water acts not only as the dispersing medium, but also as the heat
transfer medium, as well as the solvent for surfactants, initiators and, to a
lesser extent, monomer. Water quality is important, since dissolved salts
can lead to flocculation of the latex particles.

Henshaw[13] has described a typical simplified formula from which an
ionically stabilized latex may be prepared. It consists of:

Monomers (various)	100 parts
Ionic surfactants	1 part
Non-ionic surfactants	1 part
Water	100 parts
Potassium persulphate	2 parts

These components are emulsified in the reaction vessel to give a complex
system which is partly a solution and partly a dispersion.

Polymerization takes place in two broad stages. In the first of them, the
nucleation stage, the system consists primarily of droplets of dispersed
monomer, about 10 μm in size, which are stabilized by surfactant. As well as
these droplets, there are surfactant molecules and newly generated free
radicals in solution. In principle, polymerization may take place at a
number of sites, but the most likely mechanism involves the formation of
free radicals in solution, which may then develop into small growing chains
by reaction with dissolved monomer. These radicals or chains then pass
into the micelles and propagate polymerization there.

Monomer diffuses from the monomer droplets into solution, and from
there into the micelles, and propagation continues. Soon, the micelles cease
to exist as such, but instead become latex particles having surfactant
molecules adsorbed on the surface. This stage occurs at about 15–20%
polymerization in a typical vinyl acetate/acrylate system, and once it is
reached, few new particles form. The subsequent phase of the process, the
so-called 'growth' stage, proceeds by the monomer continuing to diffuse

into the latex particles from solution until the monomer droplets have all disappeared. Polymerization itself may continue to be initiated, but this happens within the latex particles, rather than promoting formation of new ones.[13]

During the overall process of emulsion polymerization, unlike bulk or solution polymerization, there is almost no increase in viscosity. Since the latex particles show no particular interaction with water, the viscosity is not found to change significantly up to about 60% solids content.[13]

The final latex needs to fulfil a number of requirements. It should exhibit good chemical stability, so that it is not destroyed by the addition of other components during paint manufacture. For the same reason, it must have good mechanical stability, i.e. the ability to withstand up to 20 min in a blender without coagulating. It should not show a substantial rise or fall in viscosity on storage, and indeed should be generally stable to storage, including freezing and thawing. Finally, it should have low foaming properties.

The final particle size of a latex varies from 0·05 to 3 μm, according to the nature of the polymer and the precise conditions of polymerization. Actual size measurement can be done using electron microscopy or by ultracentrifuging,[39] though simple visual inspection can give some indication, since latices with particles at the smaller end of the range generally have a slight blue tinge. The film-forming properties of a latex depend in part on the particle size; this overall topic of film formation by latices is dealt with in the following section.

6. EMULSION PAINTS

Paints based on pigmented latices, i.e. 'emulsion' paints, are widely used in domestic situations. They are attractive for the do-it-yourself decorator, because they are easily applied, using either brush or roller; since they dry quickly, they can be overcoated if necessary within a few hours, and tools and containers used for them can be cleaned with water.[23]

The main application for emulsion paints is decorative. They are used on interior walls and ceilings, where their matt finish and generally pale colour are highly regarded. On absorbent surfaces, they may need to be thinned slightly with water, but on non-absorbent surfaces this is not necessary and they can be used as supplied. They can be applied to a variety of substrates which occur within buildings, including plaster, concrete and timber,[39] and

16   JOHN W. NICHOLSON

they have good durability when exposed to the weather, or are in contact with alkaline substrates.

As a rule, emulsion paints dry to give fairly porous films, and for this reason they have not generally been used on metal substrates, especially ferrous ones.[15] This situation, though, is changing. The use of anticorrosive additives,[19] of novel impermeable binders,[40] or of blends of latex with water-soluble resins[15] have all been reported in recent years, and each of these developments has led to some success in preparing emulsion paints that are suitable for use on ferrous substrates.

The dried films formed by emulsion paints are generally of matt finish, gloss finishes being generally associated with the use of solvent-based systems, such as alkyds.[41] There are two reasons for this: firstly, the latex binders tend to be relatively soft, and in order for the final film to have adequate mechanical properties it has to be so heavily pigmented that the pigment particles break the film surface. Secondly, latex films are themselves inherently imperfect. They contain microvoids and other surface irregularities, and all of these detract from the natural reflectance which can be seen from the surface. There are developments here, though, including the use of much harder copolymers,[40,41] leading to reductions in the amount of pigment needed, which in turn have made it possible to formulate gloss emulsion paints much more readily.

6.1. Drying and Coalescence of Latex Films

Film formation and drying are more complex for emulsion paints than for simple solvent-based ones. In the initial stages, drying is controlled by vapour-phase diffusion as the water evaporates.[29] When this stage is complete, polymer and pigment are left behind, together with minor components of the paint formulation interspersed between them. In order for the film to be reasonably continuous the polymer particles, which are essentially spherical and separate when the film is laid down, must flatten and coalesce.[1,2] Moreover, this process must occur in such a way that the pigment particles also become fully integrated into the coating. In order for both of these to happen, the polymer must deform readily, and its ability to do so depends on two factors: (a) the natural hardness of the polymer and (b) on whether or not appropriate 'coalescing' solvents are present.[23] Since the hardness of the polymer is dependent on the temperature, as to a lesser extent is the effectiveness of any coalescing solvent, it follows that the ability of an emulsion paint to form satisfactory films is itself dependent on the prevailing temperature. Moreover, there is a minimum temperature at which film formation will occur; below this the polymer particles neither coalesce nor embody the pigment.

6.1.1. Minimum film-formation temperature

It is found in practice that the minimum film-formation temperature, MFFT, is an important property of an emulsion paint based on a synthetic latex.[42] If a film of the unpigmented latex alone is allowed to dry at an appropriate temperature, coalescence of the polymer particles causes a film to be formed that is clear and essentially continuous. By contrast, if the film is allowed to dry at a temperature below the MFFT, coalescence does not occur, and the resulting coating is white, powdery, poorly integrated and often heavily cracked.

The MFFT is a property not just of the polymer but of the entire latex formulation since it is affected by, among other things, whether or not coalescing solvents are present. These solvents are usually oxygenated ones, such as ether-esters of ethylene glycol,[23] and they are not only good solvents for the polymers in question, but are also relatively involatile. MFFT is closely related to the glass transition temperature, T_g, of the polymer. This is the temperature, or more accurately, range of temperature, at which the polymer changes from being a brittle, glassy material to a softer, plastic one due to a significant increase in the degree of segmental rotation within the polymer backbone. It is found that the factors which alter T_g also similarly affect MFFT, and hence the inclusion of 'soft' monomers in a copolymer leads to a lowering of both T_g and MFFT, whilst the inclusion of 'hard' monomers raises them.

The relationship between glass transition temperature and MFFT is not a simple one, but varies with copolymer composition. Ellgood[43] has reported the effect of copolymer composition on T_g and MFFT of various latices containing vinylidene chloride, VDC. An example made from VDC with ethyl acrylate is shown in Fig. 4. Two series of latices were prepared for this system, one dispersed with a mixed anionic and non-ionic surfactant, the other dispersed with an anionic one. The change in MFFT with composition was different in each series, thus illustrating the importance of the overall latex formulation on this property. It is of interest that the glass transition temperatures of the copolymers do not fall on a line joining the T_gs of the two homopolymers, but instead pass through a maximum at 50 mol%.[43] This is thought to be due to the inhibition of free rotation in ethyl acrylate segments by chlorine atoms on neighbouring vinylidene chloride units. Assuming an approximately random nature for the

18   JOHN W. NICHOLSON

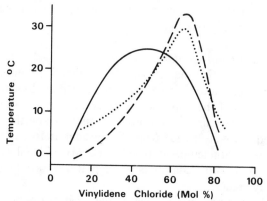

FIG. 4. Relationship of MFFT and glass transition temperature, T_g, for vinylidene chloride–ethyl acrylate copolymer: ——, T_g; - - -, MFFT (anionic/non-ionic); ....., MFFT (anionic). From Ref. 43 with kind permission of the author.

copolymer, this would be expected to be a maximum where equimolar amounts of monomer are present.

The MFFT can be seen from Fig. 4 to differ quite considerably from the glass transition temperature. At low levels of VDC, it falls below the T_g, whereas above about 55 mol% VDC, it comes above T_g. While the overall shape of the curve for MFFT is broadly similar to that for T_g, it is not symmetrical and the maximum occurs at about 65 mol% VDC. The differences between MFFT and T_g appear to be related to the hydrophilic nature of the copolymer, and this impression is confirmed since strongly hydrophilic groups, such as unsaturated acids, have an even more pronounced effect on the T_g–MFFT relationship.[43]

MFFT is generally determined by applying the coating of interest to a metal platen which can be differentially heated and cooled along its length.[42] The coating is allowed to dry on this platen and the MFFT determined by visual inspection of the resulting film. This is readily done, since the point along the surface at which the coating first forms an integrated film can be related to a particular temperature. An apparatus for carrying out this determination has recently become commercially available.[42]

6.1.2. Internal stress in emulsion paints

As a film of an emulsion paint dries, there is a build-up of internal stress.[44,45] Essentially this is because as water evaporates and the polymer particles coalesce, the coating seeks to reduce its overall energy by

shrinking. However, it is prevented from doing so by both its adhesion to the substrate, and by its own solidification. There is some stress relaxation, as the polymer molecules flow slightly and realign themselves with time, but some residual internal stress always remains. It can be demonstrated simply, by coating one side of a thin steel shim. As the coating dries, the shim deforms into a curve, which reaches a maximum radius of curvature, and then opens out slightly as the relaxation processes occur.[45]
    The development of internal stress is affected by the paint composition, i.e. pigmentation, including pigment volume concentration and nature of the filler, and type and amount of coalescing solvent. On the practical level, care has to be taken to minimize internal stress development, since otherwise spontaneous disbonding may occur; at the very least, adhesion may be severely impaired.

6.2. Polymers for 'Emulsion' Paints

The most commonly used polymers for emulsion paints are of the vinyl type.[23] The principal monomer used is vinyl acetate:

$$CH_2{=}CH.O.CO.CH_3$$

It is not generally used alone for emulsion paint binders,[2] but instead it is copolymerized with other monomers, with the exact proportions of each monomer being chosen to optimize the mechanical properties of the dried film.
    Alternatively, there are the acrylic-type binders. The parent acrylic monomer is acrylic acid, which may be regarded not only as the parent of the acrylates, i.e. esters of acrylic acid, but also of the methacrylates, via replacement of the α-hydrogen atom by a methyl group.[46] Early acrylic emulsions suffered from a number of defects, including poor wetting and flow and poor weathering resistance. Substantial improvements have since been made, so that they are now generally reckoned to be more durable than vinyl acetate copolymers, and they give better gloss.[14] This latter feature arises for two reasons: (1) acrylic resins have higher refractive indices than vinyl acetate ones, and (2) the latices have smaller particle sizes.[14] A typical acrylic emulsion paint binder is the copolymer of ethyl acrylate (70%)–acrylic acid (2·5%)–methyl acrylate (27·5%), which has a minimum film-formation temperature of 3°C.[31]
    It is, of course, possible to combine vinyl and acrylic types of monomer into a single emulsion paint binder. Thus successful coatings have been made from the polymer 2-ethylhexyl acrylate (18%)–acrylic acid (2%)–vinyl acetate (80%).[41]

TABLE 6
Contribution of Monomers to Film Properties[14]

Property	Monomer needed
Exterior durability	Acrylates and methacrylates
Hardness	Methyl methacrylate, acrylic acid, styrene
Flexibility	Ethyl or butyl acrylate, 2-ethylhexylacrylate
Water resistance	Higher methacrylates and acrylates, methyl methacrylate
Solvent and grease resistance	Acrylonitrile, methacrylamide, methacrylic acid

Various monomers are used in the preparation of synthetic latices for coatings applications, and they generally confer some distinctive property on the final copolymer. For example, butyl acrylate is used to impart flexibility, whilst styrene is used to improve hardness.[14] Table 6 lists various monomers, together with the properties to which they contribute in the final film; from this table it can be seen that it is possible to tailor copolymers to have properties appropriate for specific applications.

6.3. Formulation of Emulsion Paints

Emulsion paints are sophisticated mixtures and care is needed in their formulation. Molyneux[47] has listed up to 16 separate ingredients that may be present in a paint, though it is not likely that any one formulation would contain all of them. These include polymer particles, anionic and non-ionic surfactants, pigment and extender particles, thickener, coalescing solvent, preservative, corrosion inhibitor, antifoaming agent, antifreeze and thixotropic agent. These components together represent about half of the wet paint formulation, the remainder being water. The mixture needs to fulfil demanding requirements in terms of acceptable rheology, covering, hiding power and in-can stability, including freezing and thawing.

Emulsion paints are generally pale in colour,[23] and the main pigment used is rutile titanium dioxide.[48] The presence of the pigments means that the paints actually consist of two dispersed systems, the polymer latex and the pigment dispersion.[49] Each system requires its own dispersing agent, and it has generally been assumed that, since in each case the individual particles have negatively charged surfaces, their interaction ought to be negligible. This assumption has been shown to be doubtful by the work of Jaycock and Kayem.[49] They studied a model emulsion paint which consisted of a poly(vinyl acetate) latex with a rutile titanium dioxide pigment, surface-treated with alumina. The latex was dispersed with the aid of sodium dodecylsulphate, while the pigment was dispersed using sodium hexametaphosphate. Experiments in which the proportions of the two dispersing agents were varied revealed that the sodium hexametaphosphate tended to flocculate the latex at high concentrations. Thus it was necessary to keep its level as low as possible, consistent with stable pigment dispersion. Pigmentation is usually effected practically in two stages.[23] Initially, the pigment is dispersed simply in water with the aid of a surfactant, mixing being performed with a propeller-type disperser such as a Silversen mixer.[23] The resulting slurry is then added to the latex, and mixed further. High gloss is not generally sought in emulsion paints, and because of this, dispersion is not carried out to such a high degree as for other paints. Where extenders are used, these, too, are dispersed by this two-stage technique.

The viscosity of a simple pigmented latex is very low, and not very different from that of water.[29] In order for the paint to have sufficient 'structure' to prevent pigment settling and to give reasonable film thickness on application, it is necessary to add substances to increase the viscosity. These thickeners, which are also known as 'protective colloids', include substances such as ammonium poly(acrylate) or water-soluble cellulose derivatives, e.g. methyl- or hydroxyethyl-cellulose, or sodium carboxy-methylcellulose. These polymers, especially the cellulose derivatives, may be attacked by micro-organisms, leading either to in-can thinning of the wet paint or to mould growth on dried films. To prevent these problems, biocides are often added to the paint formulations. As mentioned earlier, there are various chemicals used for this, ranging from phenol derivatives to organometallics.[2]

6.4. Latex Paints for Corrosion Protection

The use of latex-based paints on structural steelwork is a relatively recent application, and has only been made possible by the use of novel latices. These coatings have protective, rather than decorative, functions, and this in itself is a new use for latex-based paints.

The reason that organic coatings are applied to exterior steelwork at all is to minimize or, if possible, prevent corrosion. As is well known, corrosion is the degeneration of metal by chemical or electrochemical means within the environment, and occurs because the element seeks a thermodynamically more favoured form.[6] Metals such as iron, which are classified as 'hard' in Pearson's 'hard and soft acids and bases' scheme,[50] are won from oxide ores. If they are left unprotected in the environment, they tend to revert back to their oxidized state, with all the attendant problems of loss of mechanical strength which this involves. Three principal means have been identified by which coatings may help to prevent corrosion. They are:

(1) by acting as a barrier to water or water vapour,
(2) by acting as a barrier to oxygen,
(3) by acting as a barrier to ions.

Other important aspects of their protective function include the ability to resist alkali, and the ability to retain adhesion in damp, aggressive environments.[6] Indeed, this last point has been argued to be the most important of all, and led Timmins[51] to develop the concept of 'prohesion', i.e. 'protection-is-adhesion' as the basis for the laboratory evaluation of protective paints. For corrosion to occur, two electrochemical reactions have to take place at the metal surface.[40] They are:

$$Fe \rightarrow Fe^{2+} + 2e^- \text{ (anodic reaction)}$$

$$H_2O + \tfrac{1}{2}O_2 + 2e^- \rightarrow 2OH^- \text{ (cathodic reaction)}$$

From these it can be seen that coatings which act as barriers either to water or to oxygen will inhibit corrosion by interfering with the cathodic part of the overall electrochemical reaction. In use, conventional paints are found to be quite permeable to both water and oxygen, and they derive their protective function from the high electrical resistance at the coating–substrate interface. It is this high resistance which inhibits the passage of charge that is necessary in order for the electrochemical corrosion reactions to take place.[40]

A major breakthrough in the field of protective latex paints has been the development of the 'Haloflex' series of polymers. They are the invention of the ICI Mond Division Laboratories,[40,52,53] and consist of copolymers of vinylidene chloride, vinyl chloride, and an alkyl acrylate or methacrylate, with typically a small amount of acrylic acid to confer water-dispersibility. The predominant component of these copolymers is vinylidene chloride, which is present in amounts between 65 and 95% by weight.[53] A typical copolymer composition is vinylidene chloride (69·9%)–vinyl chloride (22·4%)–2-ethylhexylacrylate (5·9%)–acrylic acid (1·8%). This copolymer, which may be prepared by emulsion polymerization at 60°C, has an MFFT of 12°C.

Early in the development of these latices, it was decided to make them highly chlorinated, in analogy to chlorinated rubber. This latter polymer, which comes in various grades, is used as the binder in solvent-borne paints.

It has been used extensively for the protection of structural steelwork, and is known to have low permeability to oxygen and to water, as well as performing well in the field. Vinylidene chloride was chosen as the base monomer because not only is it highly chlorinated, but also it is well known to confer very good barrier properties.[43]

There is one disadvantage with the use of the highly chlorinated Haloflex polymers, however, and that is their relative lack of stability. Unless care is taken, they readily undergo dehydrochlorination, i.e. loss of HCl, which introduces double bonds into the polymer backbone. This, in turn, leads to discolouration and embrittlement. It has been found in practice, though, that this is more of a problem under alkaline conditions, and that simply formulating the primers containing these latices at pH 5 makes them completely stable.

A further attraction of the Haloflex series of latices, in addition to low permeability to water and to oxygen, is that they contain a reduced amount of surfactants and protective colloids by comparison with conventional latices. Both of these groups of additives tend to increase the water-sensitivity of the final coatings, and although their use cannot be totally eliminated in practical paint formulations, they can be reduced in cases, such as this, where the parent latex does not rely on them for stability. Colloidal dispersion in Haloflex latices is maintained predominantly by the charged end groups, these groups having been introduced by the polymerization initiator during the preparation of the latex. In practical paint formulations based on Haloflex latices, the surfactants and dispersants are primarily used to maintain the pigment dispersion. However, using a latex with a very low initial surfactant content is nonetheless found to minimize the problems caused by such substances in dried films.

It is found that dried paint films produced from these latices have excellent flexibility and adhesion. Paints based on Haloflex copolymers are now becoming commercially available, and they represent a significant advance in the technology of protective coatings.[2]

6.5. Gloss Emulsion Paints
For the reasons outlined earlier, it is usual for emulsion paints to be formulated with such high pigment loadings that the dried film has a matt rather than a gloss finish. To overcome this, latices may be produced from harder polymers but the problem which then arises is that the particles of such polymers do not readily coalesce. In order for film formation to occur at ambient temperatures, it then becomes necessary to use substantial

quantities of coalescing solvent.[54] In some cases this has led to the use of such solvents at levels as high as 250 g litre^{-1}, and this clearly negates one of the prime advantages of a waterborne system.

An alternative approach is to prepare films from emulsified oligomers which are capable of cross-linking in order to develop their final hardness. This gives a low MFFT, minimizes the use of coalescing solvents and allows the formation of a film of good mechanical properties. Such an approach was adopted by Toivonen,[54] who has described an air-drying amino resin emulsion system. In this system, cross-linking occurred by an auto-oxidative mechanism similar to that in the 'drying oils' traditionally used by the paint industry.

Chemically, the copolymers used by Toivonen comprised N,N'-bis(methoxymethyl)urea and trimethylolpropane diallyl ether condensed together, with 2-ethylhexanol being used to control the dispersibility and internal plasticization of the finished resin. Oxidative 'drying' of these copolymers was aided by incorporating naphthenates of cobalt and of calcium into the paint formulations, which were then able to harden under ambient conditions to attractive, glossy films having sufficient water-resistance for indoor use. Nonetheless, they suffered from the disadvantage that the binder underwent in-can oxidation, and this led to the formation of harmful acrolein in the paints. This problem of chemical instability was not overcome by Toivonen in this particular study.[54] However, the principles he established are useful in pointing the way to gloss emulsion paints for domestic use.

7. ELECTRODEPOSITION

Electrodeposition represents one of the major uses for waterborne coatings, and is applicable to both solutions and dispersions. It is a topic of considerable complexity, and what follows is a very brief outline.

In this technique, the object to be painted is immersed in a bath of paint, and a voltage is applied between it and another electrode. Under the influence of the potential difference, paint is deposited on the article in a uniform and controlled manner. When painting is complete, the object is removed from the bath, and subjected to further treatment, usually stoving, to remove the residual water and to insolubilize the film-forming resin. The process is suitable only for metallic substrates, since it is necessary for the substrate to act as an electrode.[2]

Essentially two processes occur in the coating bath, namely electrophoresis and deposition, and the overall process is known as electrolysis.[55] In the first stage—electrophoresis—the charged particles, including ions, in solution or suspension, move in the appropriate direction in response to the externally applied field. This is followed by deposition, a term which covers the complex processes occurring at the electrodes, including the discharge of electrolytes and any subsequent reaction.

In order for the resin particles to be charged, it is necessary that the solvent should have a high dielectric constant; this limits the process of electrodeposition, for all practical purposes, to aqueous systems.[56] Some work has been done with organic solvents, but since the charges which could be obtained were much smaller than those in water, the necessary voltages were much higher.

As originally introduced into industry in the early 1960s, electrodeposition used anionic solutions and dispersions, which in general contained free or neutralized carboxylic acid groups.[57] These acid groups conferred water-solubility or water-dispersibility on the polymers, as well as providing the electrical charge necessary for electrophoresis and subsequent deposition. Since about 1978, there has been a change in the underlying technology with the introduction of cathodic electrodeposition primers.[57-59] The change in polarity provides a technique which is superior in terms of corrosion protection and throwing power, and the automobile industry has now widely adopted it.[59]

7.1. Pigmentation of Paints for Electrodeposition

The preceding discussion has been confined to the deposition of resins, but practical paints also contain pigments and extenders, and it is necessary that all of these should be deposited on to the substrate at roughly similar rates. If not, the composition of the bath will change significantly with time, and this in turn will alter the composition of the coating deposited.[56] In principle, ensuring an approximately equal rate of deposition of a mixture of two dispersions, or of a mixture of a solution and a dispersion, is formidable. However, in practice, it is found that the deposition of the resin and the pigment can be kept in step simply by correct formulation. In actual operation, there is some change in composition as deposition proceeds, because there tends to be slightly preferential deposition of the polymer, and this tends to raise the pigment:binder ratio in the bath. There is not, however, sufficient change to cause a serious drift in the composition of the paint as deposition proceeds.

The reason that pigment particles undergo electrophoresis is that they tend to adsorb polymer molecules on to their surfaces,[60] and this causes them to travel in the same direction as the paint binder when an electrical potential is applied. Because of this adsorption it is found, in general, that deposition is not affected by the grade or type of pigment.[60]

8. DEVELOPMENTS IN POLYMERS FOR WATERBORNE COATINGS

A consistent theme among the new developments currently taking place in the field of waterborne coatings is the preparation of novel film-forming polymers based on conventional materials that have been modified to confer water-solubility or dispersibility.[2] Often it is more advantageous to make such polymers dispersible rather than soluble, since this does not require such a hydrophilic polymer and, as a result, final films are less water-sensitive. On the other hand, there are a number of advantages in using coatings based on true solutions. In particular, final film gloss is superior, and there are fewer voids or other irregularities. Moreover, drying is a less complex process, and not dependent to the same extent on temperature in order to form films having good integrity. In this section some of the approaches towards polymer modification are described; in all cases, the key technique employed has been the introduction of polar groups into the polymer with the minimum of disruption to the molecule, in order to retain the desirable properties of the parent polymer in the final film.

8.1. Epoxy Resins for Can Coatings

Coatings for the interior of metal cans have traditionally been applied at very low solids in organic solvents, and it has been necessary to change this, particularly in the United States, in order to conform to the more demanding standards of pollution control.[61] In addition, it is important that these non-polluting coatings continue to meet the extremely stringent requirements necessary for food-contact applications. These include the ability to be applied by spraying or rollercoating, and the development of sufficient cross-linking in 8–10 min at 200°C or less. Finished films must be thin (2–5 μm), must impart no flavour to the contents, and must show minimal leaching of organics, none of which must be toxic.[62] In the United States, the standard for these materials is set by the Food and Drugs Administration (FDA),[62] and although the FDA regulations have no legal force in the UK, they represent the standard adopted by the canning industry in order to be able to sell to the US market. The effect of these regulations is to reduce substantially the choice of materials that can be used on the interior of metal cans.[2]

Conventionally, epoxy resins have been used in this application, and they have the following structure:

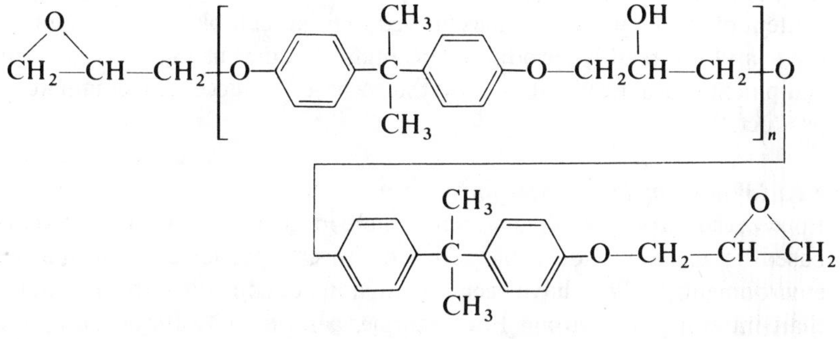

To confer water-solubility on such a polymer, one method has been to introduce carboxylic acid groups on to the aromatic rings, and then to neutralize them with volatile amines.[62,63] To introduce the carboxylic acid groups, use has been made of the Mannich reaction, i.e. the aminomethylation of phenols, as illustrated:

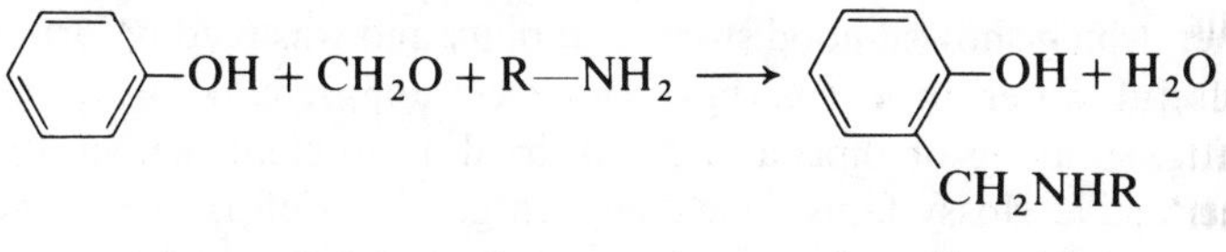

If the organic group of the amine contains a carboxylic acid group, e.g. an amino acid, this process introduces free carboxylic acid groups:

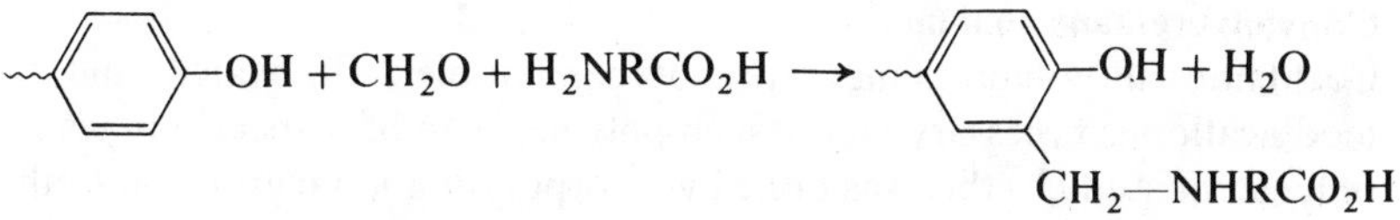

These water-soluble epoxies have been cross-linked by stoving with amino resins such as methylated melamine–formaldehyde. Cure temperatures of 200–215°C have been used for the cross-linking reaction, and the cured films were found to be suitable for the interior of beer and beverage containers, as well as for food cans. This process has several advantages, both economic and technical: in particular, the amino acids of choice are generally cheap, and the solubilization reaction is reasonably rapid.

In another approach, graft copolymers were prepared consisting of an epoxy backbone with pendant poly(acrylic acid) chains.[61] For can coatings, it is desirable to have a high level of epoxy polymer; for example, a typical resin of this kind would have an epoxy content of 80% and an acrylic acid content of 20%. Such a resin has proved to be acceptable for food-contact uses, and is readily produced in high volume using conventional equipment; as a result, it is now the basis of a successful commercial product.[61]

8.1.1. Water-soluble epoxies for other uses
Epoxy resins are versatile materials which, in their conventional, solvent-based form, are widely used in coatings,[64] particularly in marine environments.[65] They have been prepared in aqueous form for uses other than interior can coating. For example, a series of waterborne epoxy coatings for exterior steelwork were prepared using fatty acids from dehydrated castor oil.[66] They were heated with maleic anhydride at 200–210°C for 6 h under nitrogen; this was followed by further modification, for example by reaction with allyl alcohol, to increase unsaturation and to develop 'drying' properties. These maleinized fatty acids were then used to partly esterify epoxy resins, yielding a complex polymer, which showed good storage stability and was readily formulated into useful waterborne coatings. These coatings could be applied by brushing, spraying or dipcoating, and dried at ambient temperatures to give attractive glossy films which showed good anticorrosive properties. Furthermore, they could be combined as both primer and topcoat in a single system.[66]

8.2. Polyurethane Ionomers
Conventional polyurethanes are not compatible with water, and so modification is necessary before such polymers can be dispersed in water. In a recent study,[67] this was done by incorporating ionic groups into the polymer backbone. These ionic groups, being hydrophilic, acted as internal emulsifying agents, to make the polymers self-dispersing. The term 'ionomer' has not been used for such systems before. It was originally used by DuPont for their range of lightly carboxylated hydrocarbon and fluorocarbon polymers,[68] but the term has become appropriate for any polymer system containing a minority of ionized functional groups, usually 10% or less.[68] In the recent work, both cationic and anionic polyurethane

ionomers have been prepared. The cationic types were prepared by including tertiary amine groups in the polymer, whilst the anionic ionomers were made from carboxylated monomers. Dimethylpropionic acid was used extensively, and neutralized with a tertiary amine. In a third approach, polar polyether chain segments were included in the polymer, yielding an internally emulsified polyurethane which was dispersed non-ionically.

As they stand, such polyurethane dispersions form coatings having properties which are inferior to conventional solvent-borne two-pack polyurethanes. However, the inclusion of potential cross-linking reagents into these formulations can effect significant improvements. For example, alkoxylated melamine–formaldehyde, which can itself form stable aqueous dispersions, was incorporated into the mixture, and a film was formed · which was cured by stoving. Cross-linking took place by the reaction of methoxylated melamine with urea or urethane groups in the parent polymer. These water-dispersed cross-linkable polyurethanes gave very good cured films, with properties approaching those of conventional solvent-borne systems.[67]

8.3. Bunte Salts as Solubilizing Groups

Carboxylation is not the only method that has been used to confer water-solubility. Recently, Thames[69] has described a new method based on the formation of Bunte salts. These are substituted aminoethanethiosulphuric acids, AETSAs, and contain the functional groups

$$-NH-CH_2-CH_2-S-\overset{\displaystyle O}{\underset{\displaystyle O}{\overset{\|}{\underset{\|}{S}}}}-OH$$

As the functional group in a polymer, this structure confers water-solubility, and is then capable of thermal disproportionation and cross-linking. Adequate cross-links have been found to form at temperatures as low as 123°C, and the resulting films are not water-sensitive. The cross-links actually consist of sulphur bridges, as illustrated:

$$-CH_2-S-S-CH_2-$$

There are a number of routes to Bunte salts. For example, they can be made by the reaction of alkyl halides with sodium thiosulphate in aqueous solutions at fairly high temperatures:

$$R-X + Na_2SSO_3 \rightarrow R-SSO_3^- Na^+ + NaX$$

Alternatively, a displacement reaction with aminoethanethiosulphuric acid itself is possible:

$$R-X + H_2N-CH_2-CH_2S-SO_3H \rightarrow R-NH-CH_2-CH_2S-SO_3H$$

This process has been used to solubilize poly(epichlorohydrin), the reaction being carried out by refluxing in a 5:1 dimethylformamide:water mixture at 85°C for 48 h. Although a similar treatment of poly(vinyl chloride) would appear to be attractive, in practice the polymer simply underwent dehydrochlorination under reflux conditions.

The water-soluble or dispersed polymers produced from poly(epichlorohydrin) were found to be stable in water for long periods of time with the addition of external emulsifiers. They underwent uncatalysed cross-linking at 123–135°C to form films which showed excellent adhesion to metal substrates, and excellent resistance to the solvent methyl ethyl ketone.[69]

9. CROSS-LINKING REACTIONS

Efficient cross-linking is often crucial since in almost every application of waterborne coatings maximum water-resistance is demanded. Fortunately, however, the functional groups that confer water-solubility or -dispersibility are often reactive enough to be the sites at which cross-linking can occur.

One well-established cross-linking reaction is that of amide-N-methylol ether groups either with themselves or with hydroxyl groups. This reaction, which requires elevated temperatures, has been used to cross-link electrodeposited coatings.[2] The two possible reactions are

$$-CO.NH.CH_2OH + HO- \rightarrow -CO.NH.CH_2-O- + H_2O$$

and

$$-CO.NH.CH_2OH + HOCH_2.NH.CO-$$
$$\rightarrow -CO.NH.CH_2.NH.CO- + H_2O + CH_2O$$

In addition to such 'self' cross-linking reactions, various 'external' cross-linking reactions have been used, often with amino or phenolic precondensates as the cross-linking agent. Such cross-linkers typically contain methylol or alkoxy groups, and these groups react with hydroxyl groups on the parent resins, liberating water, and yielding stable cross-links. Good examples of these curing agents are the derivatives of hexamethoxymethylmelamine, HMMM, which has the structure

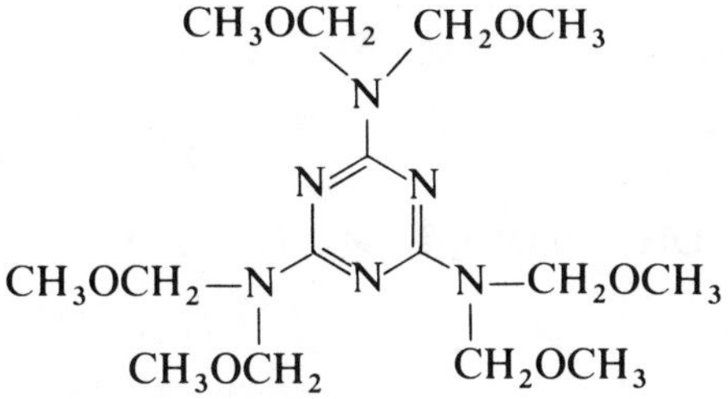

$$CH_3OCH_2 \quad CH_2OCH_3$$

Partial reaction of HMMM with maleic anhydride gives a product which is water-soluble, and can be used in anodic electrodeposition.[2] Like the main resin, it migrates to the anode when current is passed and so is deposited in the initial film. On stoving, it reacts with the carboxylic acid groups of the resin to give water-resistant cured films.

9.1. Latent Acid Catalysts For Cross-linking
The widely used condensation reactions for curing waterborne coatings are generally acid-catalysed.[2] An attractive way of carrying out such a reaction is to make use of a latent acid catalyst.[70] These catalysts are latent in the sense that they are not acidic at ambient temperature and are not reactive, and hence have good storage stability. At higher temperatures, however, they generate sufficient acid to bring about extensive cure. Typical latent acid catalysts are sulphamate esters, which at high temperatures rearrange to their betaine forms, which then undergo hydrolysis in water to yield acid:

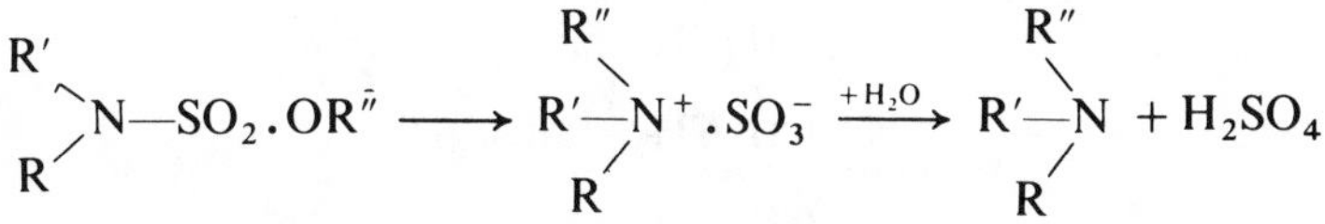

The amine, being volatile, is lost from the system, leaving the sulphuric acid behind to catalyse the cross-linking reaction.

This overall rearrangement will occur at temperatures as low as 121°C, and may in future prove useful for reducing the cure temperature of a number of similar coatings that are cured with melamine or its derivatives.

9.2. Titanium Esters as Cross-linking Agents
Tetra-alkyl titanates and their derivatives are useful cross-linking agents, and have been used in the preparation of cured coatings of various

polymers.[71] They readily undergo two reactions that are especially useful for waterborne coatings, namely ester interchange and reaction with carboxylic acids:

Ester interchange:

$$2 \times \{\!\!-OH + Ti(OR)_4 \longrightarrow \{\!\!-O-\overset{\displaystyle OR}{\underset{\displaystyle OR}{\overset{|}{\underset{|}{Ti}}}}-O-\!\!\} + 2ROH$$

Reaction with carboxylic acids:

$$2 \times \{\!\!-C\overset{\displaystyle O}{\underset{\displaystyle OH}{\diagup}} + Ti(OR)_4 \longrightarrow \{\!\!-O-\overset{\displaystyle OR}{\underset{\displaystyle OR}{\overset{|}{\underset{|}{Ti}}}}-O-\!\!\} + 2ROH$$

Titanium also has a tendency to expand its coordination number to six by forming chelates with available carbonyl groups, for example on carboxylated polymers, and this also contributes to the cross-linking and water-insolubility of the coatings.

Among the derivatives which have been used is the triethanolamine isopropoxytitanate chelate. This, in its commercially available form, consists of a mixture of chelated species, one of which has the cage structure[71]

$$N\begin{matrix} \diagup CH_2CH_2 \diagdown \\ -CH_2CH_2- \\ \diagdown CH_2CH_2 \diagup \end{matrix}TiOC_3H_7$$

It has been used in the cure of water-soluble alkyd resins,[72] giving hard, water-resistant coatings suitable as maintenance enamels for use on farm implements, for example.[72]

9.3. Zirconium Compounds as Cross-linking Agents
Another recent development in the field of cross-linking agents has been the use of zirconium compounds for carboxylated resins that are water-soluble.[73] The particular zirconium compounds used tend to form

condensed polymeric structures in aqueous solution. For example, zirconyl chloride at pH 6 has the following structure:

$$
\begin{array}{ccccc}
\text{OH} & & \text{OH} & & \text{OH} \\
| & \text{OH} & | & \text{OH} & | \\
\text{Zr} & & \text{Zr} & & \text{Zr} \\
+ & \text{OH} & + & \text{OH} & + \\
\text{Cl}^- & & \text{Cl}^- & & \text{Cl}^-
\end{array}
$$

Similar species exist in alkaline solution, such as in ammonium zirconium carbonate. These zirconium compounds are able to interact with the functional groups of organic polymers, forming either relatively weak hydrogen bonds with hydroxyl groups or strong covalent bonds with carboxylic acid groups.

A typical film cured by this means consisted of a latex of butyl acrylate, methyl acrylate and acrylonitrile with 3% acrylic acid as the 'internal' emulsifying agent. The film was cured with zirconium acetate at 130°C for 30 min, after which it showed excellent resistance to acetone.[73] By comparison a film of the latex alone, without any added zirconium salt, when subjected to the same treatment, showed almost no resistance to acetone.

One of the attractions of this approach is that zirconium is non-toxic, and a cross-linking agent of this type, ammonium zirconium carbonate, has recently been given FDA approval for use in food-contact applications.[73]

10. FURTHER DEVELOPMENTS IN WATERBORNE COATINGS TECHNOLOGY

In addition to the improvements already described in the areas of water-solubilization and cross-linking, there have been other developments in the technology of waterborne coatings; these are discussed in this section.

10.1. Waterborne Coil Coating

Precoated metal coil is an increasingly important commercial material,[7] having quadrupled in output in the 15-year period from 1969, to reach a level of about 200 000 tons a year worldwide by the mid-1980s. The reason for the growth in this manufacturing technique is that it is easier, and hence cheaper, to precoat large areas of metal sheet, and then stamp out the individual components after forming. The metal substrates employed are usually either steel or aluminium. The coatings need to survive demanding mechanical requirements, as the sheet is coiled, uncoiled, cut, formed and assembled, and there are a number of solvent-based lacquers which are able to do so.

Increasingly, waterborne coatings have been developed for this application. The most successful have been those based on either polyester or acrylic resins, cross-linked with a water-soluble melamine derivative during stoving. The polyesters used have consisted of a mixture of adipic and isophthalic acids, together with trimellitic anhydride, esterified with an excess of polyfunctional hydroxy compounds, such as pentaerythritol, glycerol or trimethylolpropane. The acrylics have been copolymers of various alkyl acrylates, acrylic acid, hydroxyacrylates or hydroxymethacrylates, together with methacrylic acid and styrene. This latter monomer is included in order to reduce the cost of the final coating.

In each case, the complex copolymers contained substantial numbers of hydroxyl groups, and, in the case of the acrylics, carboxylic acid groups as well. These are the sites at which cross-linking has taken place, usually with HMMM or its derivatives. The resulting films have been chemically complex, and this has enabled them to stand up well to the rigorous demands put upon them in the coiling and subsequent manufacturing processes.

10.2. Water-dispersed Powder Coatings

These represent a combination of two environmentally acceptable technologies, namely waterborne and powder coatings. In principle, they combine the ecological advantages of powder coatings with the ease of application of conventional wet paints.[74]

Powder coatings are usually applied by electrostatic spraying. The powder itself consists of finely divided, essentially uncross-linked polymer, often an epoxy resin. When the coating has been applied, it is heated and this causes it to flow out to form a continuous, consolidated film. It also brings about cross-linking.

As a technique, this is of growing importance, but it does have certain disadvantages. It is difficult to produce thin films, high cure temperatures are required, and the overspray, which can be collected for re-use, has a different particle size distribution from the original powder. Some of these problems can be overcome if the powder is prepared as a dispersion in water. For instance, it is easier to prepare thin films, and less material falls into the overspray region. There is little difference between the performance of powders applied conventionally and those applied as an aqueous dispersion, except for a slight discolouration of the latter on ferrous substrates due to flash rusting. There were no reported disadvantages with this particular system,[74] though the presence of residual surfactant, needed originally to stabilize the dispersion, might be expected to cause problems under certain circumstances.

10.3. Flash-Rusting Inhibitors

As mentioned in Section 2, a potentially serious problem with waterborne coatings is flash rusting on ferrous substrates. It may be overcome by using suitable inhibitors, though care is needed in their selection in order to prevent deleterious side effects, such as water-sensitivity or salt-spray fogging.[19] The New England Society for Coatings Technology carried out a major study of potential inhibitors, in the course of which they identified a number of useful compounds.[19] For example, ammonium chromate, ammonium dichromate and sodium dichromate were found to give excellent protection against flash rusting, but imparted a yellow colour to the films. Of the numerous inorganic salts examined, the best were found to be sodium nitrate, sodium nitrite and potassium nitrite. All gave good resistance to flash rusting and caused only very slight salt-spray fogging and water-sensitization. A number of organic compounds were examined, including tributylamine, N-aminoethylethanolamine and metal carboxylates, such as calcium octoate and sodium benzoate. The nitrogen-containing molecules gave good results, whilst the carboxylates gave rather variable protection against flash rusting. For example, the octoates of calcium or lead were good, with little in the way of fogging or water-sensitization, whereas other carboxylates such as sodium acetate or sodium benzoate improved resistance to flash rusting, but gave very serious fogging and showed substantial water-sensitivity. Altogether, a large number of effective additives were identified in this study but, being based on an entirely empirical approach, it was not clear by what mechanism(s) any of them worked.

10.4. Waterborne Printing Inks

Recently, there has been significant progress in the development of waterborne inks for both printing and packaging applications,[75] though they have actually made far less impact on the industry than waterborne paints.

Among the earliest waterborne inks were those for newspapers designed for use with rotary letterpress machines. There were problems, however, with premature drying, and tacky films causing paper fibres to be deposited on the plate. More recently, emulsion inks consisting of aqueous dispersions of mineral oils pigmented with carbon black have been made, and these appear to have better properties. They require good stability to the shear forces that occur in the printing processes; when this is achieved, they retain the good printability of conventional oil-based inks. One of the main problems in the area of inks for packaging applications has been the need to print on impermeable substrates, such as plastic film. It is found that adhesion and water-resistance of aqueous inks are inferior to those of the best solvent-based ones. For example, a waterborne ink prepared from an alkali-soluble acrylic copolymer and pigmented with Phthalocyanine Blue was used on a polyethylene film. Although on drying the ink gave good scratch resistance and adhered well to the polyethylene, after soaking in water for 24 h both scratch resistance and adhesion were severely reduced. Such water-sensitivity would be unacceptable, for example, in the low-temperature conditions typically used for storing frozen pre-packed foods.

Overall, developments are occurring with waterborne inks, but progress is slow. The most promising areas for such inks are with relatively absorbent substrates, namely paper and board, rather than impermeable ones such as polyethylene, which continue to present difficult problems.[75]

11. CONCLUSION

In this chapter, an attempt has been made to identify the important principles that underly the preparation and use of waterborne coatings, and to deal with some of the many developments in this field which have been made in recent years. As a class, waterborne coatings are continuing to grow in commercial importance. Moreover, in view of the environmental acceptability and cost-effectiveness of using water as a paint solvent, there is every prospect that, for the forseeable future, this growth will be sustained. Technical innovation, especially in the areas of water-solubilization and cross-linking, will be the key to progress.

REFERENCES

1. MARTENS, C. R., *Waterborne Coatings*, 1981, van Reinhold Company, New York.
2. NICHOLSON, J. W., *Waterborne Coatings*, Oil and Colour Chemists' Association Monograph No. 2, 1985, Oil and Colour Chemists' Association, London.
3. VALENTINE, L., *J. Oil Colour Chemists Assoc.*, 1984, **67**, 157.
4. FETTIS, G. C., *J. Oil Colour Chemists Assoc.*, 1985, **68**, 159.
5. FINCH, C. A., *Chem. & Ind. (London)*, 1981, 800.

6. NEW ENGLAND SOCIETY FOR COATINGS TECHNOLOGY, *J. Coatings Technol.*, 1981, **53**(683), 27.
7. PERCY, E. J. and NOUWENS, F., *J. Oil Colour Chemists Assoc.*, 1979, **62**, 392.
8. TYLER, T. J. C., *J. Oil Colour Chemists Assoc.*, 1983, **66**, 17.
9. DAVIES, D. S. and LAWRENSON, I. J., *J. Oil Colour Chemists Assoc.*, 1979, **62**, 68.
10. BROWN, R. A., *J. Coatings Technol.*, 1983, **55**(707), 75.
11. VAN WESTRENAN, W. J., *J. Oil Colour Chemists Assoc.*, 1979, **62**, 246.
12. LUTHARDT, H. J. and BURKHARDT, W., *J. Oil Colour Chemists Assoc.*, 1979, **62**, 375.
13. HENSHAW, B. C., *J. Coatings Technol.*, 1985, **57**(725), 73.
14. PAUL, S., *Progr. Org. Coatings*, 1977, **5**, 79.
15. NAKAYAMA, Y., WATANABE, T. and TOYMOTO, I., *J. Coatings Technol.*, 1984, **56**(716), 73.
16. DUFOUR, P., *J. Oil Colour Chemists Assoc.*, 1979, **69**, 59.
17. MYERS, R. R., *J. Macromol. Sci.—Chem.*, 1981, **A15**(6), 1133.
18. ANDERFERATI, F. B., *J. Oil Colour Chemists Assoc.*, 1980, **63**, 367.
19. NEW ENGLAND SOCIETY FOR COATINGS TECHNOLOGY, *J. Coatings Technol.*, 1982, **54**(684), 63.
20. WINDIBANK, B. P. and MAHAR, R. W., *J. Oil Colour Chemists Assoc.*, 1979, **62**, 426.
21. PAUL, S., *Surface Coatings*, 1985, John Wiley and Sons, Chichester.
22. BOXALL, J. and WORLEY, W., *J. Oil Colour Chemists Assoc.*, 1979, **62**, 173.
23. MORGANS, W. M., *Outlines of Paint Technology*, Vol. 2, 2nd edn, 1984, Charles Griffin and Co., High Wycombe.
24. RISBERG, M., *J. Oil Colour Chemists Assoc.*, 1985, **68**, 197.
25. ARNOLDUS, R., *J. Oil Colour Chemists Assoc.*, 1985, **68**, 263.
26. SHAW, D. J., *Introduction to Colloid and Surface Chemistry*, 2nd edn, 1970, Butterworths, London.
27. SCHOFF, C. K. and PIERCE, P. E., in *Organic Coatings*, Vol. 7, ed. G. D. Parfitt and A. V. Patsis, 1984, Marcel Dekker, New York.
28. FRANKS, F., *Water*, 1983, Royal Society of Chemistry, London.
29. KORNUM, L. O., *J. Oil Colour Chemists Assoc.*, 1980, **63**, 103.
30. SPEAKMAN, J. C., *The Hydrogen Bond*, 1975, The Chemical Society, London.
31. GILMORE, G. N., *A Modern Approach to Comprehensive Chemistry*, 1975, Stanley Thomas (Publishers), London.
32. ROBB, I. D., in *The Chemistry and Technology of Water-Soluble Polymers*, ed. C. A. Finch, 1983, Plenum Press, New York.
33. HUGHES, L. J. T. and FORDYCE, D. B., *J. Polym. Sci.*, 1956, **22**, 509.
34. FRANKS, F., in *The Chemistry and Technology of Water-Soluble Polymers*, ed. C. A. Finch, 1983, Plenum Press, New York.
35. MORRISON, R. T. and BOYD, R. N., *Organic Chemistry*, 3rd edn., 1973, Allyn and Bacon, New York.
36. SILCOCKS, C. G., *Physical Chemistry*, 1966, McDonald and Evans, London.
37. BATZER, H. and LOHSE, F., *Introduction to Macromolecular Chemistry*, 2nd edn, 1979, John Wiley and Sons, Chichester.
38. VAN WIEL, H. and ZOM, W., *J. Oil Colour Chemists Assoc.*, 1981, **64**, 263.
39. VORSTER, O. C., *J. Oil Colour Chemists Assoc.*, 1979, **62**, 299.
40. BURGESS, A. J., CALDWELL, D. and PADGET, J. C., *J. Oil Colour Chemists Assoc.*, 1981, **64**, 175.

41. MERCURIO, A., KRONBERGER, K. and FRIEL, J., *J. Oil Colour Chemists Assoc.*, 1982, **65**, 227.
42. GORDON, P. G., DAVIES, M. A. S. and WATERS, J. A., *J. Oil Colour Chemists Assoc.*, 1984, **67**, 197.
43. ELLGOOD, B., *J. Oil Colour Chemists Assoc.*, 1985, **68**, 164.
44. PERERA, D. Y., *J. Coatings Technol.*, 1984, **56**(716), 111.
45. PERERA, D. Y., *J. Oil Colour Chemists Assoc.*, 1985, **68**, 275.
46. BRYDSON, J. A., *Plastics Materials*, 4th edn, 1982, Butterworth Scientific, London.
47. MOLYNEUX, P., in *The Chemistry and Technology of Water-Soluble Polymers*, ed. C. A. Finch, 1983, Plenum Press, New York.
48. SIMPSON, L. A., *J. Coatings Technol.*, 1984, **56**(715), 57.
49. JAYCOCK, M. J. and KAYEM, G. J., *J. Oil Colour Chemists Assoc.*, 1982, **65**, 431.
50. PEARSON, R. G., *J. Amer. Chem. Soc.*, 1963, **85**, 3533.
51. TIMMINS, F. D., *J. Oil Colour Chemists Assoc.*, 1979, **62**, 131.
52. BOXALL, J., *Polym. Paint & Col. J.*, 1984, **174**(4125), 578.
53. BURGESS, A. J. and LAWRENCE, D. (ICI Ltd), Ger. Offen. 2 756 000, 6 July 1978; *Chem. Abstr.*, **89**, 131197.
54. TOIVONEN, H., *J. Oil Colour Chemists Assoc.*, 1984, **67**, 213.
55. TAWN, A. R. H. and BERRY, J. R., *J. Oil Colour Chemists Assoc.*, 1965, **48**, 790.
56. FINN, S. R. and MELL, C. C., *J. Oil Colour Chemists Assoc.*, 1964, **47**, 219.
57. SCHENCK, H. U., SPOOR, H. and MARX, M., *Prog. Org. Coatings*, 1979, **7**, 1.
58. SCHENCK, H. U. and STOELTING, J., *J. Oil Colour Chemists Assoc.*, 1980, **63**, 482.
59. QUICK, H. L., *J. Oil Colour Chemists Assoc.*, 1983, **66**, 361.
60. ENTWISTLE, T., *J. Oil Colour Chemists Assoc.*, 1972, **55**, 480.
61. ROBINSON, P. V., *J. Coatings Technol.*, 1981, **53**(674), 23.
62. DEMMER, C. G. and MOSS, N. S., *J. Oil Colour Chemists Assoc.*, 1982, **65**, 249.
63. DEMMER, C. G. and MOSS, N. S., *Polym. Paint & Col. J.*, 1982, **172**(4077), 524.
64. POTTER, W. G., *Epoxide Resins*, 1970, Illiffe Books, London.
65. BANFIELD, T. A., *Marine Finishes*, Oil and Colour Chemists Association Monograph No. 1, 1980, Oil and Colour Chemists' Association, London.
66. SHIRSALKAR, M. M. and SIVASAMBAN, M. A., *J. Oil Colour Chemists Assoc.*, 1982, **65**, 301.
67. TIRPAK, R. E. and MARKUSH, P. H., Paper presented at the *Twelfth Waterborne and High Solids Symposium*, University of Southern Mississippi, New Orleans, 13–15 February, 1985. Quoted in *Waterborne and High Solids Bulletin*, 1985, **7**(9), 1.
68. LONGWORTH, R., in *Developments in Ionic Polymers*, Vol. 1, eds A. D. Wilson and H. J. Prosser, 1983, Applied Science Publishers, London.
69. THAMES, S. F., *J. Coatings Technol.*, 1983, **55**(706), 33.
70. HART, D. J., *J. Coatings Technol.*, 1983, **55**(704), 87.
71. ANON., *Polyfunctional Tyzor® Organics Titanates*, Technical literature from DuPont, Chemicals and Pigments Dept, Wilmington, Delaware 19898, USA.
72. LERMAN, M. A., *J. Coatings Technol.*, 1976, **48**(623), 37.
73. MOLES, P. J. and BYRAM, J., *Polym. Paint & Col. J.*, 1984, **174**(4122), 440.
74. NORTH, A. G., *J. Oil Colour Chemists Assoc.*, 1981, **64**, 355.
75. HUTCHINSON, G. H., *J. Oil Colour Chemists Assoc.*, 1985, **68**, 306.

CHAPTER 2

Electrodeposition of Paints

H.-J. Streitberger

BASF Lacke und Farben AG, Münster, Federal Republic of Germany

and

R. P. Osterloh

BASF AG, Ludwigshafen, Federal Republic of Germany

1. INTRODUCTION

The electrocoating of electroconductive substrates is a well-established technology of industrial painting.[1,2] Under the influence of a direct current, charged polymers migrate electrophoretically to the electrode (workpiece) of opposite charge. At the electrode the polymers become insoluble and form an insulating film. Technical development based on this phenomenon began early in this century. In 1919 a patent[3] was granted for painting of conductive articles. Later some applications were described for electrodeposition of latices, oleoresinous layers and additional protection of the inner surfaces of tin-plated cans.[4,5] The industrial importance of the electrodeposition process was emphasised when the process was adapted for car-body priming in the beginning of the 1960s. Brewer and Burnside[6,7] developed this technology together with paint manufacturers, firstly using the anodic process (AED) which grew to form nearly 100% of the priming technology in Europe and Japan's automotive industry as well as 50% in the United States. By the end of the 1970s, a changeover to cathodic electrodeposition (CED) started so that by 1985 almost all electrodeposition was by this process.[8,9]

The industrial success is based on a number of advantages which CED offers:

(1)   Superior corrosion resistance at low film thicknesses and in small hollow spaces (high throwing power), good edge protection and uniform coating thickness.
(2)   High coating utilisation (>95%).
(3)   Application process easy to automate and to control.
(4)   Closed loop process possible.
(5)   Low level of pollution (aqueous system).
(6)   Easy transfer of AED process to CED and vice versa possible.

Intensive research in chemical companies, the coating industry and the paint-consuming industry led to significant improvements in paint quality and especially to a better understanding of fundamental aspects of AED and CED.[10-17]

It is the aim of this chapter to give an overview of the recent developments in chemistry, physical aspects and technology in CED.

2.   CHEMISTRY OF BINDERS FOR CED

2.1. General Considerations

Resins for CED are polymers which bear positive charges along and/or at the end of their polymer chains or they are neutral polymers kept in solution by cationic emulsifiers. Water-dispersibility of the polycations is achieved by complete or partial neutralisation with a water-soluble carboxylic acid leading to the formation of a polysalt. Acetic and lactic acids are widely used for this purpose (Fig. 1).

Pigments, water-insoluble solvents, neutral polymers and curing agents are encapsulated by these polysalts. The polycations migrate under the

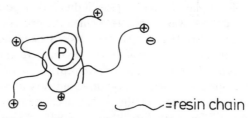

FIG. 1.   Partially neutralised, positively charged resins encapsulating pigment particles P.

influence of the electric field to the cathode where they are discharged and deposited.

The uncured film adheres to the electroconductive substrate and soon builds up a high resistance which slows down further deposition. Today, commercially used CED-binders are thermosetting resins of low relative molar mass (M_n 2000–15 000). Thermosetting resins are preferred because the crosslinking reaction offers a range of possibilities:[18]

(1) To obtain an insoluble film after baking with tailored mechanical film properties by varying the degree of crosslinking.

(2) To block or eliminate ionic groups which are essential for electrodeposition but lead to problems of poor corrosion resistance of the film.

The structural possibilities for binders and crosslinking units or self-crosslinked binders are covered by nearly 1200 patents. There are very many different polymers including polycondensates, (aldehydes, amino and phenolic plastics, polyamides), polyaddition polymers (polyepoxides, polyurethanes) and polymerised resins (polyvinyl compounds, (meth)acrylates and modified-dienehomo- and co-polymers), but all have three common functions: they can be

(1) rendered soluble;
(2) made elastomeric;
(3) crosslinked.

Where a high level of corrosion resistance is required, such as in the automotive industry, only epoxy-based resins have been used to date.

2.2. Solubilisation

Resin modification to confer aqueous solubility has been achieved with a variety of primary, secondary and tertiary amine groups as well as quaternary ammonium salts or quaternary ammonium hydroxides,[19] ternary sulphonium acid salts[20,21] and quaternary phosphonium acid salts.[22] However, only nitrogen-containing groups have gained industrial importance for this purpose.

2.2.1. Amino-containing polymers

Radical copolymerisation of (meth)acrylic esters, styrene, vinyl ethers and vinyl esters with unsaturated monomers containing sterically hindered secondary amino groups or tertiary amino groups (e.g. aminoalkylamides of (meth)acrylic acid, mono- or di-alkylaminomethyl-substituted styrenes, vinylimidazole) leads in one step to cationic polymers.[23-29] Groups capable of conferring elastomeric properties and crosslinking can easily be introduced at the same polymerisation step, so even self-crosslinkable polymers may be prepared via copolymerisation with unsaturated, blocked N-alkenylisocyanates[30-32] or di-isocyanates mutually blocked with hydroxyalkyl esters of (meth)acrylic acid and primary or secondary alcohols[33] or with esters of methylol-(meth)acrylamides.[34]

2.2.2. Epoxy-based resins

Epoxy groups containing copolymers (e.g. copolymers of glycidyl methacrylate)[35,36] or bisphenol-A epichlorhydrin adducts are reacted with ammonia, amines or ammonium salts of organic acids.

$$\sim\!O\!-\!CH_2\!-\!CH\!-\!CH_2 + NH_3 \longrightarrow \qquad\qquad \text{Refs 37, 38}$$
$$\underset{O}{\diagdown\diagup}$$

$$\sim\!O\!-\!CH_2\!-\!\underset{\underset{OH}{|}}{CH}\!-\!CH_2\!-\!NH_2$$

$$\sim\!O\!-\!CH_2\!-\!CH\!-\!CH_2 + NH_2\!-\!R \longrightarrow \qquad\qquad \text{Refs 39 41}$$
$$\underset{O}{\diagdown\diagup}$$

$$\sim\!O\!-\!CH_2\!-\!\underset{\underset{OH}{|}}{CH}\!-\!CH_2\!-\!NH\!-\!R$$

$$\sim\!O\!-\!CH_2\!-\!CH\!-\!CH_2 + NH(R)_2 \longrightarrow \qquad\qquad \text{Refs 42, 43}$$
$$\underset{O}{\diagdown\diagup}$$

$$\sim\!O\!-\!CH_2\!-\!\underset{\underset{OH}{|}}{CH}\!-\!CH_2\!-\!N(R)_2$$

$$\sim\!O\!-\!CH_2\!-\!CH\!-\!CH_2 + N(R)_3 + CH_3COOH \longrightarrow \qquad \text{Refs 44 46}$$
$$\underset{O}{\diagdown\diagup}$$

$$\sim\!O\!-\!CH_2\!-\!\underset{\underset{OH}{|}}{CH}\!-\!CH_2\!-\!N^+(R)_3 + \cdot CH_3COO^-$$

In addition, several ways are known for introducing primary and secondary amino groups using excess of primary amine, diamine or polyamine,[47] imidazolidines,[48] oxazolidines[49] or other capped primary amines (e.g. ketimines of primary/secondary amines or primary diamines).[50-53]

$$\sim\!O\!-\!CH_2\!-\!\underset{\underset{O}{\diagdown\diagup}}{CH}\!-\!CH_2 + NH_2\!-\!R\!-\!NH_2 \longrightarrow$$

$$\sim\!O\!-\!CH_2\!-\!\underset{\underset{OH}{|}}{CH}\!-\!CH_2\!-\!NH\!-\!R\!-\!NH_2$$

$$\sim\!O\!-\!CH_2\!-\!\underset{\underset{O}{\diagdown\diagup}}{CH}\!-\!CH_2 + HN\!-\!CH_2 \longrightarrow$$

(ring amine: HN—CH_2, R_1R_2C, CH_2, N—H)

$$\sim\!O\!-\!CH_2\!-\!\underset{\underset{OH}{|}}{CH}\!-\!CH_2\!-\!N\!-\!CH_2$$

(ring: R_1R_2C, CH_2, N—H)

$$\downarrow +H_2O$$

$$\sim\!O\!-\!CH_2\!-\!\underset{\underset{OH}{|}}{CH}\!-\!CH_2\!-\!NH\!-\!(CH_2)_2\!-\!NH_2 + \underset{R_2}{\overset{R_1}{\diagdown}}C\!=\!O$$

$$\sim\!O\!-\!CH_2\!-\!\underset{\underset{O}{\diagdown\diagup}}{CH}\!-\!CH_2 + R_1\!-\!CH\!-\!CH_2 \longrightarrow$$

(with HN, O, C, R_2, R_3)

$$\sim\!O\!-\!CH_2\!-\!\underset{\underset{OH}{|}}{CH}\!-\!CH_2\!-\!N\ \ (R_1\!-\!CH\!-\!CH_2,\ O,\ C,\ R_2\ R_3)$$

$$\overset{[H^+]}{\longrightarrow}\downarrow +H_2O$$

$$\sim\!O\!-\!CH_2\!-\!\underset{\underset{OH}{|}}{CH}\!-\!CH_2\!-\!NH\!-\!\underset{\underset{R_1}{|}}{CH}\!-\!CH_2OH + \underset{R_3}{\overset{R_2}{\diagdown}}C\!=\!O$$

$$\text{\textasciitilde\hspace{-2pt}O—CH}_2\text{—CH—CH}_2 + HN \overset{\displaystyle R_1\text{—N=C}(R_2R_3)}{\underset{\displaystyle R_1\text{—N=C}(R_2R_3)}{\diagdown}} \longrightarrow$$

$$\text{\textasciitilde\hspace{-2pt}O—CH}_2\text{—}\underset{\underset{\displaystyle OH}{|}}{\text{CH}}\text{—CH}_2\text{—N} \overset{\displaystyle R_1\text{—N=C}(R_2R_3)}{\underset{\displaystyle R_1\text{—N=C}(R_2R_3)}{\diagdown}}$$

$$[\text{H}^+]\big\downarrow + \text{H}_2\text{O}$$

$$\text{\textasciitilde\hspace{-2pt}O—CH}_2\text{—}\underset{\underset{\displaystyle OH}{|}}{\text{CH}}\text{—CH}_2\text{—N} \overset{\displaystyle R_1\text{—NH}_2}{\underset{\displaystyle R_1\text{—NH}_2}{\diagdown}} + 2 \overset{\displaystyle R_2}{\underset{\displaystyle R_3}{\diagdown}}\text{C=O}$$

$$\text{\textasciitilde\hspace{-2pt}O—CH}_2\text{—CH—CH}_2 + (R_2R_3)\text{C=N—R}_1\text{—N=}(R_2R_3) \longrightarrow$$

$$\text{\textasciitilde\hspace{-2pt}O—CH}_2\text{—CH—CH}_2$$

$$[\text{H}^+]\big\downarrow + \text{H}_2\text{O}$$

$$\text{\textasciitilde\hspace{-2pt}O—CH}_2\text{—}\underset{\underset{\displaystyle OH}{|}}{\text{CH}}\text{—CH}_2\text{—NH—R}_1\text{—NH}_2 + 2 \overset{\displaystyle R_2}{\underset{\displaystyle R_3}{\diagdown}}\text{C=O}$$

These reactions lead to CED-binders with bath pH values of 6·5–7·5, good coating properties and especially low corrosivity towards the equipment. Dispersion stability and film properties of the uncured as well as the cured film are strongly influenced by the molar mass of the polyepoxide and of the type of epoxy-resin chain enlargement. Often mixtures of blocked and unblocked amines (e.g. alkanolamines) are used. The primary and secondary amino groups as well as the alcohol groups not only contribute to solubility, but at the same time they are crosslinkable with capped polyisocyanate or aminoplastic precondensates, which can be based on melamine, benzoguanamine or urea, and activated esters.

Quaternary ammonium salts are thermally unstable. During the cure process—above temperatures of 100°C—they decompose to allyl esters which undergo consecutive crosslinking reactions.

$$\sim\!\!O-CH_2-CH-CH_2-\overset{+}{N}(R)_3\cdot OH^- \xrightarrow{\Delta T}$$

$$\underset{OH}{|}$$

$$\sim\!\!O-CH_2-CH=CH_2 + H_2O + N(R)_3$$

A third aspect of the resin modification with polar groups like amino or quaternary ammonium groups is that they assist the wetting of pigments, which must be co-dispersed with the CED-binder. CED-binders, especially those modified with tertiary amine or quaternary ammonium groups, are milled together with organic solvents, pigments and extenders, catalysts and organic acids using ball or sand mills in order to optimise the wetting.

2.2.3. Amino-modified epoxy resins

Besides the direct addition of amines to epoxy resins, several routes are known to introduce amino functions into epoxy-based polymers: for example, hydroxyl groups containing epoxy resins can be reacted with partially blocked isocyanates. Dialkylalkanolamines,[54,55] N-hydroxy-alkyl-substituted oxazolidines or ketimine-blocked alkanolamines[58] have been used as blocking agents.

$$\sim\!\!O-(CH_2)_3-OH + OCN-R_1-NCO + HO-(CH_2)_n-N\overset{R_2}{\underset{R_3}{<}}$$

$$\downarrow$$

$$\sim\!\!O-(CH_2)_3-O-\overset{O}{\overset{\|}{C}}-NH-R_1-NH-\overset{O}{\overset{\|}{C}}-O-(CH_2)_n-N\overset{R_2}{\underset{R_3}{<}}$$

$$\sim\!\!O-(CH_2)_3-OH + OCN-R_1-NCO + HO-(CH_2)_2-\underset{\underset{CH_2CH_2}{|\quad|}}{N-CH_2}$$

$$\downarrow$$

$$\sim\!\!O-(CH_2)_3-O-\overset{O}{\overset{\|}{C}}-NH-R_1-NH-\overset{O}{\overset{\|}{C}}-O-(CH_2)_2-\underset{CH_2\quad CH_2}{N-CH_2}$$

$$\sim\!O\!-\!(CH_2)_3\!-\!OH + OCN\!-\!R_1\!-\!NCO + HO(CH_2)_n\!-\!N\!=\!C\!\overset{R_2}{\underset{R_3}{\diagdown}}$$

$$\downarrow$$

$$\sim\!O\!-\!(CH_2)_3\!-\!O\!-\!\overset{O}{\overset{\|}{C}}\!-\!NH\!-\!R_1\!-\!NH\!-\!\overset{O}{\overset{\|}{C}}\!-\!O\!-\!(CH_2)_n\!-\!N\!=\!C\!\overset{R_2}{\underset{R_3}{\diagdown}}$$

$$[H^+]\ \Big\downarrow\ +H_2O$$

$$\sim\!O\!-\!(CH_2)_3\!-\!O\!-\!\overset{O}{\overset{\|}{C}}\!-\!NH\!-\!R_1\!-\!NH\!-\!\overset{O}{\overset{\|}{C}}\!-\!O\!-\!(CH_2)_n\!-\!NH_2 + \overset{R_2}{\underset{R_3}{\diagup}}C\!=\!O$$

Epoxide-phenolic resins with terminal phenolic groups are able to react with mixtures of formaldehyde and secondary amines or else polyphenols are modified by condensation with formaldehyde and secondary amines and added to epoxy resins.[59-62]

$$-\!\!\langle\bigcirc\rangle\!-\!OH + CH_2O + HN(R_1R_2) \longrightarrow -\!\!\langle\bigcirc\rangle\!-\!OH \qquad + H_2O$$
$$\underset{CH_2-N(R_1R_2)}{}$$

In a similar way, (meth)acrylamide-containing copolymers can be modified by condensation with formaldehyde and secondary amines.[63]

$$\sim\!\overset{O}{\overset{\|}{C}}\!-\!NH_2 + CH_2O + HN(R_1R_2) \longrightarrow$$

$$\sim\!\overset{O}{\overset{\|}{C}}\!-\!NH\!-\!CH_2\!-\!N(R_1R_2) + H_2O$$

Since these methylene-amino groups are unstable at elevated temperatures, they contribute to solubility as well as to crosslinking reactions by splitting off secondary amines.[64] Tertiary amino groups can be introduced into polymers by the addition of secondary amines to activated double bonds.[65] Before adding to epoxy resins, polyphenols are reacted with formaldehyde and (meth)acrylamide and in a consecutive step the amine is added:

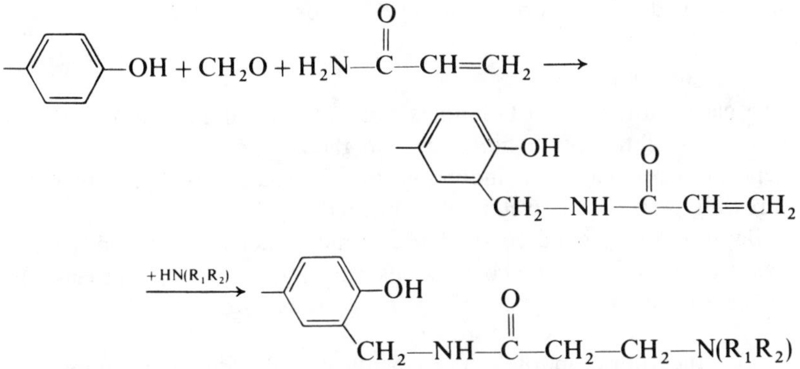

2.2.4. Amino group introduction to maleic anhydride-containing resins

Maleinized polybutadiene oils preferably with a molar mass of 1000–1500 and containing 60–70% of 1,2-vinyl groups or copolymers of maleic anhydride, vinyl compounds and esters of (meth)acrylic acid can be modified by amidation with a primary/tertiary diamine.[66,67]

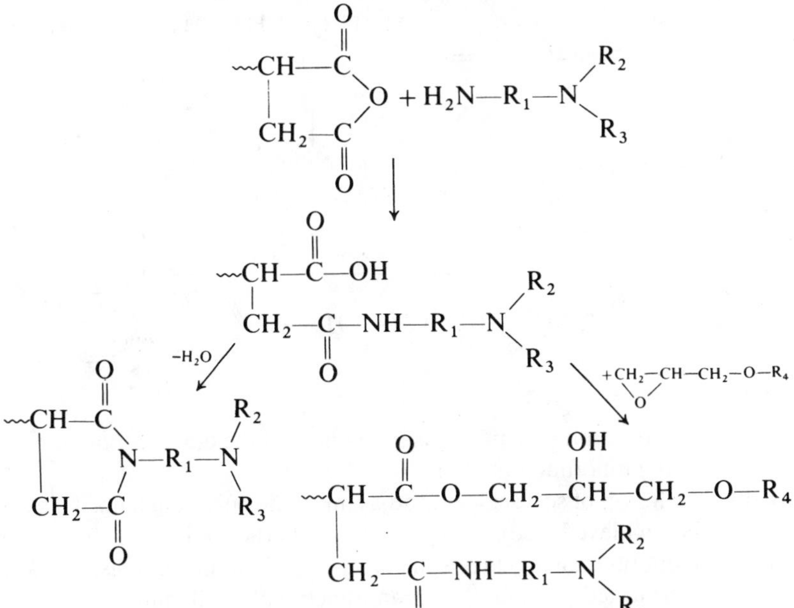

In a consecutive reaction step the remaining carboxylic acid group may be converted by thermal treatment to a substituted imide or with a monoepoxide to the corresponding β-hydroxyalkyl ester.

2.3. Elastification

The elastification of CED-binders not only determines the mechanical properties of the cured film, but also the dispersion stability, electrodeposition behaviour (i.e. induction time for coagulation, film thickness, throwing power) and levelling of the cured film.

Because a very good corrosion resistance is achieved by epoxy resins based on bisphenol-A, much work has been done to improve the elasticity of these resins. Three types of elastification are used:

(a) Incorporation of soft segments (mainly aliphatic compounds, molar mass 100–1500, such as dialcohols,[68,69] polyether- and polyester-diols,[70-73] dicarbonic acids, disecondary amines or amidamines,[74-77] diphenylolalkylcarbonylic esters[78] and modified polybutadienes[79-83] in the polymer chain.

The reaction scheme is idealised. Parts of the diol compound react only monofunctionally.

(b) Inclusion of soft segments adjacent to the polymer chain. This can be achieved along or at ends of the polymer chains by monofunctional carboxylic acids,[84-88] monoalcohols,[89] alkyl-substituted phenols[90] or monofunctional amidamines.[91]

$$2 \sim\!\!O-\!\!\langle\bigcirc\rangle\!\!-\!\!\langle\bigcirc\rangle\!\!-O-CH_2-CH-CH_2 + HO-R$$
$$\downarrow \quad \underset{O}{\diagdown\!\diagup}$$

$$\sim\!\!O-\!\!\langle\bigcirc\rangle\!\!-\!\!\langle\bigcirc\rangle\!\!-O-CH_2-\underset{OR}{CH}-CH_2$$

$$O-CH_2-\underset{OH}{CH}-CH_2-O-\!\!\langle\bigcirc\rangle\!\!-\!\!\langle\bigcirc\rangle\!\!-O\sim$$

(c) Addition of plasticisers with reactive groups, such as hydroxyl-containing acrylics[92,93] or polyethers, which are incorporated in the film during curing reaction. Usually they act as plasticisers as well as flow agents.

2.4 Crosslinking

CED-binders need to have mainly hydroxyl-, amino- or double-bond-containing groups if crosslinked with a separate crosslinker. This may be achieved by blocked polyisocyanates,[94-96] polyesters,[97] activated polyesters like β-hydroxyalkyl or carbalkoxyalkyl esters,[98-102] phenol-formaldehyde–amine adducts,[103] multifunctional crosslinkers with dienophilic end groups, which may react with diene end groups of the CED-binder,[104,105] polyfunctional unsaturated crosslinkers as prepared from the reaction of a di-isocyanate half-blocked with a hydroxyalkylacrylic ester and added to a polyol.[106]

In addition, crosslinkers may be prepared by condensation reactions of urea, diamines and secondary monoamines with polyols.[107]

Amino and phenolic resins are also described as crosslinkers;[108] however, because the transetherification must be catalysed by strong acids such resins are only of importance for AED-crosslinking, and to date they have not proved very successful for CED. Therefore, mixtures of resins and blocked polyisocyanates are used.[109] Some self-crosslinking CED-binders have been described previously.[66,67] Others can be prepared by reacting half-blocked di-isocyanates with hydroxyl groups of the binder resin.[110] Other blocking groups have been cited and include allylamines,[111] hydroxylalkylacrylic esters[112] or diene/dienophilic compounds like furfuryl alcohol and allyl alcohol.[113]

3. PHYSICAL ASPECTS OF ELECTRODEPOSITION PROCESSES

3.1. Characteristics of Electrodepositable Dispersions

To provide electrodepositable paints it is necessary to make ionically stabilised dispersions of binders (see Section 2.2). Because of energy and stability considerations, the size and surface charge density must be within a certain range.[10,114]

Resin molecules form micelles of different particle size dependent on the number of ionised salt groups per unit volume of insoluble resin phase. According to Debye,[115] a diameter of 120 nm is preferable, as has been found by light-scattering measurements.[114] Table 1 shows that the typical particle sizes are not very much different for cationic and anionic resins and that the degree of neutralisation, i.e. the amount of electrical charge, is also similar.

The larger the particle sizes are, at the same degree of neutralisation, the lower is the dispersion stability. On the other hand, the smaller the particle size, the more electrical energy is necessary for the deposition process. The relation between the number of salt groups per unit volume of resin and particle size is described in Fig. 2 for anionic resins.

All of these basic considerations for anionic paints are also valid for a cationic stabilised aqueous dispersion.

3.2. Electrodeposition Process

The mechanism of the electrodeposition process can be divided into three steps. In the first stage a boundary layer is established at the electrode surface to provide electrodeposition conditions. The second step is the kinetic growth of the film and finally an equilibrium between coagulation and dissolution is reached limiting film growth. The first stage has been investigated by many authors.[10,11,14-17,114] Fundamental studies by

TABLE 1
Typical Physical Data for Resins used in Electrodeposition Processes[10,114]

Resin	$MEQ_a{}^a$	$MEQ_b{}^b$	Total neutralisation (%)	M_w	Particle size (nm)
Anionic	0·4–2·0	—	40–90	1 000–20 000	10–100
Cationic	—	0·4–1·5	30–70	1 000–20 000	50–250

^{a} Milliequivalents of KOH/g solid.
^{b} Milliequivalents of HCl/g solid.

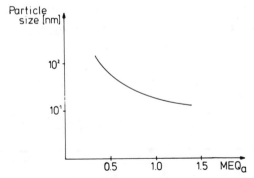

FIG. 2. Relation between resin particle size and degree of neutralisation for an anionic paint.[114]

Beck,[10,16] using a rotating disc, showed that a certain time is required for establishing a diffusion boundary layer at the metal surface before electrodeposition of a coherent film can occur.

Under the normal conditions employed in the electrodeposition process, i.e. more than 80 V and a current density of more than 1 mA/cm^2, the main reactions at the anode and cathode are dissolution of water:

Anode reaction: $2H_2O \longrightarrow O_2 + 4H^+ + 4e^-$ (1)

Cathode reaction: $4H_2O + 4e^- \longrightarrow 2H_2 + 4OH^-$ (2)

Total reaction: $H_2O \longrightarrow H_2 + \frac{1}{2}O_2$

The cathode reaction creates double the gas volume in comparison with the anode reaction under the same electrical conditions. This phenomenon seems to play an important part in determining the electrical resistance of the deposited film.

Different side reactions due to different electrode materials can occur. Anderson[117] studied the metal dissolution behaviour of anodic and cathodic electrodeposition materials (Fig. 3). For steel, the inorganic conversion layer of zinc or iron phosphates influences the dissolution rates.

Spoor[17] also found that the influence of metal dissolution on the coagulation behaviour between anionic and cationic polymers was different at different electrode surfaces (Table 2).

All these various studies demonstrate that, for the important substrates steel and galvanised steel, electrode side reactions of the anodic process contribute to the deposition process and thus influence the film properties more than the cathode reactions. Aluminium is a more suitable substrate

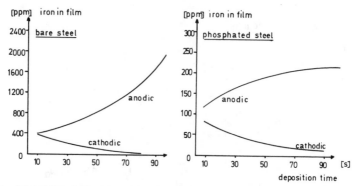

FIG. 3. Metal dissolution for an anodic and cathodic electropainting process on bare and phosphated steel.[117]

for the anodic process because of the formation of an oxide layer during the electrodeposition process.

The concentration of OH^- or H^+ ions in the boundary layer of the cathode or the anode, neglecting any electrode side reaction, can be calculated by the combination of Faraday's law and Fick's diffusion laws in eqns (3) and (4):

$$C_{OH^-} = \frac{2j}{F}\left(\frac{t}{\pi D_{OH^-}}\right)^{1/2} [1/cm^3] \tag{3}$$

$$C_{H^+} = \frac{2j}{F}\left(\frac{t}{\pi D_{H^+}}\right)^{1/2} [1/cm^3] \tag{4}$$

where F is the Faraday constant, j is the current density and D the diffusion coefficient for OH^- ions or H^+ ions. According to Nernst, a quasi-stagnant

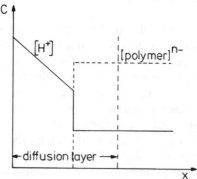

FIG. 4. Model of diffusion layer for protons and anionic paint at the anode.[118]

TABLE 2
Metal Content of Electrodeposited Films Formed by Cathodic and Anodic
Processes in Comparison with Coulombic Yields[17]

Substrate	Anodic		Cathodic	
	Metal content of film (%)	Coulombic yield (mg/C)	Metal content of film (%)	Coulombic yield (mg/C)
Fe	0·05	10	0·015	4·5
Al	0·037	14·3	0·200	3·2
Ni	0·37	10·3	0·002	4·2
Cu	1·3	7	0·046	4·1
Zn	—	—	0·094	4·0
Fe + 10 mM NaCl	1·5	17	—	—
Conditions	2 min/100 V/pH 7·7		1 min/140 V	

diffusion layer is established at a phase boundary in a stirred bath. The
mass transport in this layer is provided by diffusion and migration
processes. A concentration gradient drives a diffusion flux and also a
migration flux. Both forces can be described[118] by eqn (5):

$$J_{H^+,OH^-} = 2D_{H^+,OH^-} \frac{dc}{dx} \tag{5}$$

At the same time, paint micelles move in the direction of the electrode
surface creating a concentration gradient analogous to Fig. 4. Some time
elapses before the critical OH^- or H^+ ion concentration in the boundary
layer is reached, at which the coagulation processes arising from the
neutralisation of the polymer micelle occur [reactions (6) and (7)].

$$RCOO^- + H^+ \rightleftharpoons RCOOH\downarrow \tag{6}$$

$$R_1\overset{+}{\underset{R_2}{N}}\overset{H}{\underset{R_3}{}} + OH^- \rightleftharpoons R_2-\overset{R_1}{\underset{R_3}{N}}\downarrow + H_2O \tag{7}$$

The product of current density j and the square root of the induction time τ
is a constant.

$$j\sqrt{\tau} = K[A\,s^{1/2}/cm^2] \tag{8}$$

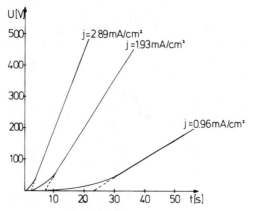

FIG. 5. Electrodeposition of oil-modified polyester on to a zinc phosphated steel substrate at different constant current densities.[115]

The condition for the beginning of deposition is then:

$$\tau = \frac{K^2}{j^2}\,[s] \tag{9}$$

Experiments carried out with constant current density demonstrate the induction time of the deposition process (Fig. 5).[23]

A plot of current density versus the reciprocal square root of the induction time (Fig. 6) demonstrates the validity of eqn (9) and the applicability of Sand's equation, in which the relationship of τ to current

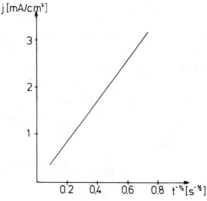

FIG. 6. Plot of current density against reciprocal square root of induction time.[116]

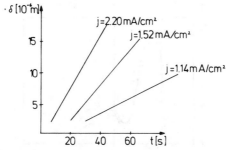

FIG. 7. Film thickness versus time for galvanostatic experiment at various current densities.[116]

density is described.[116] The slope is a constant characteristic of the electrodeposited material.

Most of the studies regarding kinetics of film growth were made by galvanostatic (constant current) experiments. After the induction time τ for establishing the diffusion layer, the electrode potential and the amount of deposited film increase rapidly. This rise is normally linear (Fig. 7) and Pierce[116] describes the film growth according to the theory of oxide film growth on metals.[11]

$$\frac{d\delta}{dt} = C(j\text{-}j_D) \text{ [cm/s]} \tag{10}$$

where $d\delta/dt$ is the rate of film growth, C is the coulombic efficiency (cm³/A s), j is the measured current density (A/cm²) and j_D is the current density at which no dissolution of deposited film occurs.

A similar equation is obtained by using Faraday's law, i.e. that the mass (m) of the film should be proportional to the charge Q (A s):

$$m = C_g Q \text{ [g]} \tag{11}$$

where C_g is the coulombic efficiency in g/A s. Differentiation of eqn (11) and using $m = \rho \delta f$ with ρ = specific density, δ = film thickness and f = area leads to eqn (12):

$$\frac{d\delta}{dt} = \frac{C_g}{\rho} j \text{ [cm/s]} \tag{12}$$

with

$$\frac{C_g}{\rho} = C \Rightarrow \frac{d\delta}{dt} = Cj \text{ [cm/s]} \tag{13}$$

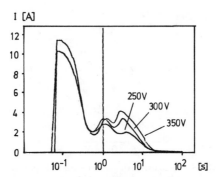

FIG. 8.   Plot of current versus time of potentiostatic electrodeposition experiments
with a cationic electrocoating.[123]

Figure 5 demonstrates the ohmic behaviour and validity of Faraday's
laws after a certain induction time even though many experiments show
non-ohmic behaviour of electrodeposition paints.[6,119,120] An explanation
for these facts may be reached by considering the build-up of charge
barriers at the beginning of coagulation within the deposited film.[121,122]
Industrial electrodeposition processes are normally conducted under
constant voltage conditions. Because of the very rapid build-up of the
diffusion layer, coagulation processes begin within a few milliseconds, as
shown by studies carried out more recently than the original galvanostatic
experiments.[123] A typical I versus t plot (Fig. 8) shows the rapid decrease of
I in the first 10 s.
A film thickness versus time plot under potentiostatic conditions looks
different from those for galvanostatic experiments (see Figs 7 and 9).
Attempts at a mathematical description of these phenomena were made
by Pierce[116] based on integration of eqn (10) and defining j_D as the limiting
factor in film build-up.

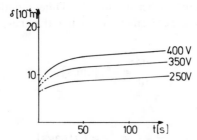

FIG. 9.   Film thickness versus time under galvanostatic conditions.[116]

3.3. Throwing Power

A very important aspect of the technical electrodeposition process is the ability to paint inner areas of complicated structures, such as automotive bodies. Many experiments and test methods have been carried out to compare the throwing power of different industrial paint systems, and attempts have been made to measure the electrical conditions of the electrodeposition process of certain test boxes.[124]

In addition, many workers have tried to calculate the application behaviour of electrodeposited paint on simple test objects.[125-128] Film build-up can be calculated from experimentally developed models quite adequately using laboratory-scale test methods to assess throwing power (Fig. 10).[128] These calculations are based predominantly on specific resistivity and time curves provided by separate application experiments.

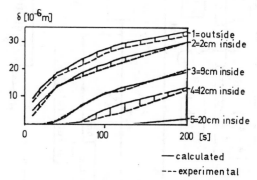

FIG. 10.   Comparison of calculated and experimental film thickness based on the throwing power test.[128]

4.   THE TECHNOLOGY OF ELECTRODEPOSITION PROCESSES FOR INDUSTRIAL USE

4.1. Introduction: General Considerations

From a general standpoint there are five main factors which influence the technology of electrodeposition processes:

(1)   Whether the process is to be anodic or cathodic.
(2)   The paint system.
(3)   The shape and dimensions of parts to be coated.
(4)   Productivity.
(5)   Costs.

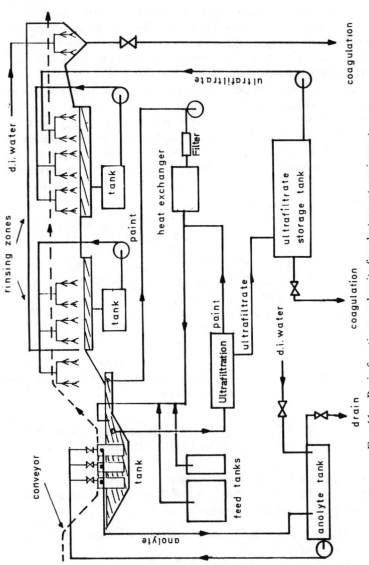

FIG. 11. Basic functions and units for electrocoat equipment.

The differences between plants suitable for anodic or cathodic electro-deposition processes are mainly found in the electrical insulation and material problems caused by low pH values of most cathodic electropaints. The way of feeding an electrocoat tank is independent of the electrodeposition process, i.e. whether it is anodic or cathodic, but has some influence on the equipment design because of differences in compensation technology. This will be pointed out in more detail in Section 4.2. Of course shape, size and type of parts which have to be coated substantially influence the size and design of electrocoat equipment.

Productivity, i.e. jobs per hour, will affect the tank design considerably. Mostly, for small parts, truck cabins and very low productvity demands, a vertical tank design is suitable. Such a system is the cheapest way of electrocoating parts because it demands less space and less equipment—for example, only one rectifier. Tank sizes for this specific use range from 0·5 tonne to 80 tonnes paint content. Passenger cars, wheels, appliances and many other industrial products are coated in equipment with running conveyors. Tank sizes for these uses range from 10 tonnes up to 500 tonnes. Electrical conditions have to be carefully considered so that these types of tanks normally have more than two rectifiers. Transfer efficiency of the electropaint is the dominant factor controlling the running costs. Use of the technology of ultrafiltration and reverse osmosis adds to the efficiency and reduces the running costs.[129–131]

The basic functional units for electrocoat equipment are a dip tank, an ultrafiltration unit, rinsing zones, anode or cathode material or equipment, a drain tank, a heat exchanger and a compensation unit (Fig. 11).

The most highly developed technology for passenger car priming with cathodic electrocoatings will be discussed in subsequent sections.

4.2. Compensation Technology

Dispersions of organic compounds such as resins in water have two kinds of stability problem. Both the physical stability and the chemical stability of those systems suitable for electrodeposition processes have to be optimised (see Section 3.1).

The reduction in chemical stability, for example of polybutadiene resin to oxygen and of polyethers to saponification by water at extremes of pH, leads to physical instability of the dispersions. This aspect is one of the basic demands on resin chemists when developing electrocoat resins. On the other hand, the stability of the physical state strongly influences the reliability of the painting process. This stability will also be stressed by

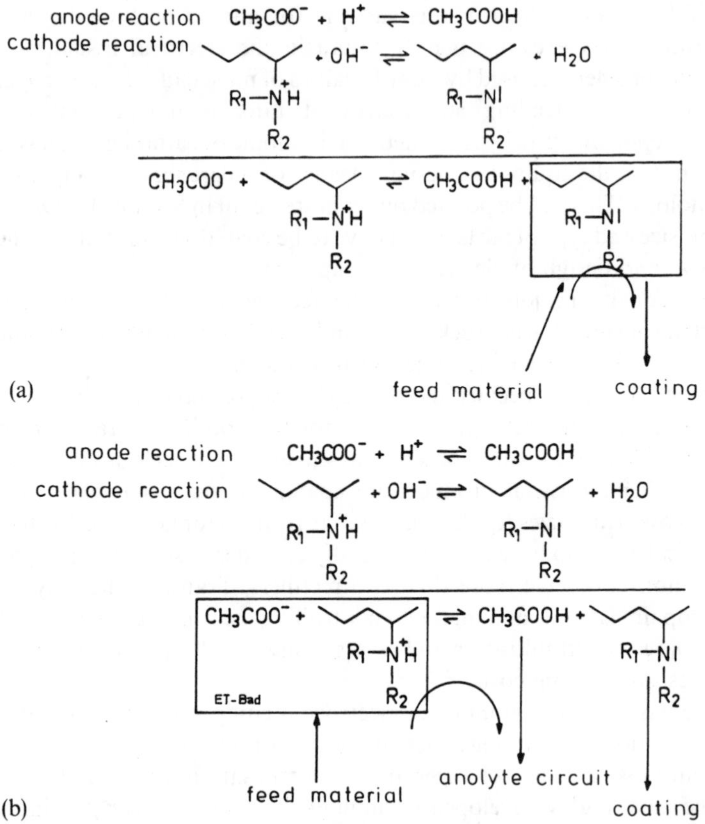

FIG. 12. Comparison of compensation technologies: (a) underneutralised compensation material; (b) completely neutralised material.

the equilibrium which will be established by the painting and compensation process.

Today two compensation techniques are in use (Fig. 12). The first is operated with underneutralized materials, mostly one-component and used in general industry. This technology, depicted in Fig. 11, requires excess neutralisation agent in the bath to dissolve the paint material. This excess has to be generated between each compensation process because paint is removed by the electrocoating process. So the condition of the paint is wave-like, corresponding to the extremes of high neutralisation, i.e. when there is excess neutralisation agent, and minimal neutralisation at the time shortly after the compensation process.

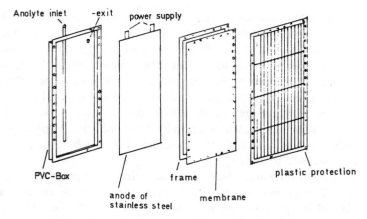

FIG. 13. Typical construction of an anolyte box.

The second technique operates with completely neutralised compensation materials. This is mostly used in the automotive industry as a two-[51] or one-[56,57] pack system and requires an anolyte system in a cathodic paint process. This anolyte system regulates by means of an electrically controlled anionic exchange membrane the concentration of acid and so keeps the paint in the same condition (Fig. 13).[132] In contrast to the first technology there is no need for dispersing units. It is sufficient to pump the feed material directly from the drums into the paint circulation system. Ion-exchange membranes are more useful in anolyte systems than simple semipermeable membranes because they operate like an electrolyte barrier outside of the operation time and only let through anions during the operation. Attempts have been made to control pH by removing the neutralisation agent (lactic or acetic acid for CED) from the ultrafiltrate circuit.[133]

Both technologies have been in use for anodic and cathodic paint processes using different types of membrane in the case of completely neutralised materials. The anolyte and catholyte circuits are precisely controlled by on-line measurement of the specific conductivity. An alternative control method utilises the osmotic pressure which is built up during the electrodeposition process.[134]

4.3. Ultrafiltration and Reverse Osmosis Techniques

The first electrocoat lines operated with a simple water rinse area so that the recovery of paint adhering to the parts was very low. In addition to the economic loss the rinse water created a potential pollution problem. These problems have been overcome by the introduction of ultrafiltration units into the paint process.[129-131,135] The transfer efficiency of the electrocoat paint could be increased to more than 95%, which significantly improves the commercial viability of the electrocoat process. Also, waste water can be much more easily handled because it is reduced to less than 1 litre/m² of coated area. Furthermore, the ultrafiltration unit operates like a kidney. Replacement of the ultrafiltrate drain by deionised water helps to reduce impurities such as salts and other detrimental materials carried into the electropaint tank by the objects.

The ultrafiltration unit is fed by a separate circulation pump (see Fig. 11). The ultrafiltrate is pumped into the last rinsing area and is normally used in a counterflow spray or dip rinse system for the electrocoated parts. Depending on the type of unit, it is useful to have a feed and bleed system for minimising energy costs. Tubular systems with diameters of 3–5 cm are characterised by a high paint flow rate per membrane area. Other low-energy units have lower rates but are more sensitive to the paint systems and respond more strongly to the physical conditions of the paint. The membranes normally are semipermeable, with separation limits of relative molar mass between 10 000 and 50 000. The ultrafiltrate therefore contains water, salts, solvents and low molar mass compounds of the paint. Characteristic data for an ultrafiltrate are given in Table 3.

The membranes must be stable towards all the different solvents which remain in an electropaint. Charging the surface of the inner membrane area by cations or anions lengthens the lifetime of the units. Otherwise cleaning-up procedures have to be carried out frequently because the paints have the tendency to deposit on the membrane surface. This reduces the ultrafiltrate flux drastically.

TABLE 3
Comparison of Typical Cathodic Electropaint and its Ultrafiltrate

	Solids[a] (%)	pH	Specific conductivity (mS/cm at 20°C)	Solvents (%)
Electropaint	20	6·2	1·2	3·0
Ultrafiltrate	0·4	5·9	0·8	1·2

[a] 2h/130°C.

Increased environmental demands in the last few years make it sometimes necessary to have a completely closed-loop system for the electrocoating process. Under these circumstances a reverse osmosis unit can be installed in the ultrafiltrate circuit.[131]

4.4. Electrical Power Supply

For the commercially used electrodeposition process direct current is necessary: it is normally applied by thyristor-controlled rectifiers. The fluctuation of the direct current has to be as low as possible (less than 5%) to avoid cratering and rough surfaces on the paint film. Depending on the electrical resistivity of the paint and electrode distances the voltages are in the range from 200 to 500 V. The current is also dependent on the paint system and on the area which has to be coated. It can reach more than 1500 A/rectifier for a big automotive tank (Table 4).

A vertical dip tank is the easiest way of providing the electrical power supply. Only one rectifier is necessary to form the desired current versus time curve for the coating process. More complicated are the conveyor lines where the influence of geometrical conditions, size and shape of parts as well as line speed on the current have to be considered. In such lines it is sometimes difficult to achieve a uniform film thickness on sections of very different geometry.

To avoid excessively high current densities at the beginning of the electrodeposition process, and thereby cratering,[136] it is necessary to install at least two rectifiers. Thus it is possible to provide a defined voltage programme which consists of a lower voltage for the tank entry area and standard voltage for the rest of the tank.

For anodic electrodeposition processes the total tank is sometimes the cathode, which is then earthed. Otherwise some unprotected steel panels are used to provide a separate controlled cathode area. The cathodic

TABLE 4
Typical Electrical Data for Electrodeposition Equipment

Voltage	150–450 V
Current	800–2 000 A/rectifier
Anode/cathode ratio[a]	1:4 to 1:6
Anode–cathode distance[b]	> 30 cm
Current density	4–6 m A/cm^2

[a] For cathodic process.
[b] For body tanks.

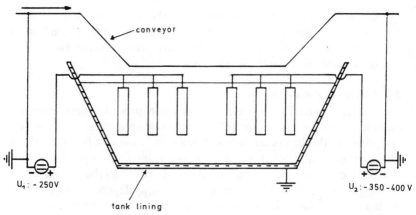

conveyor

tank lining

U_1 : ~ 250V

U_2 : ~ 350 - 400 V

FIG. 14.   Simple circuit for cathodic electrodeposition power supply.

process requires special anode material such as stainless steel or graphite-containing polymers which will not dissolve too quickly. To avoid dissolution and dirt producing coagulation of paint, the tanks for cathodic processes are lined with highly resistant organic coatings. The parts and the tank are earthed and the anodes are held at a positive potential (see Fig. 14). To obtain separate voltage zones, it is useful to install a gap between the entry anodes supplied by the first rectifier and the next anodes. In Europe a

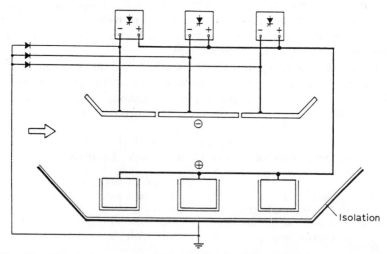

Isolation

FIG. 15.   Electrical power supply for a cathodic electrodeposition process with a divided conveyor.

defined power control system can often be found, for example by dividing the conveyor into three parts. In the case shown in Fig. 15, there is a separate contact on the conveyor system, so that potential is applied between the conveyor contact device and the anode. Also the tank is earthed and diodes are used to prevent an electrical breakdown which might happen when some of the lining of a tank has been destroyed. In every case all rectifiers must be designed to give a slow build-up of the voltage. This is necessary to avoid excessively high currents at the beginning of the painting process. Normally it takes some time to optimise the electrical conditions for every unit which at present cannot be calculated completely.

5. COMMERCIAL USE

The most important industrial use for the electrodeposition process is undercoating car bodies for corrosion protection and as a primer for subsequent topcoating.

To achieve good adhesion as well as improving corrosion protection, the total car body has to be pretreated to obtain an inorganic conversion layer; the pretreatment is normally a zinc phosphating of the steel surface.[137] Since the beginning of the 1980s these conversion layers have also needed to be suitable for galvanised steel, which has grown in use for heavy-corrosion protection in car construction.[138]

Adhesion is greatly dependent on both the crystal structure of the surface and the alkaline resistivity of the phosphate salt, so the proportion of phosphophylite to hopeite crystals on steel is important for good corrosion resistance.[139] (Table 5).

Shortly after introduction of the electrodeposition process to the

TABLE 5
Development of Electrodeposition Technology in the Automotive Industry

	1962	1966	1973	1977	1982
Application method	Spraying/dipping	Electro-coating	Electro-coating	Electro-coating	Electro-coating
Film thickness (mm)	0·030–0·050	0·025–0·030	0·020–0·025	0·015–0·020	0·025–0·035
Solvent content[a] (%)	100	80	70	20	10
Resin chemistry	Alkyd	Maleinised oils	Maleinised polybutadienes	Epoxy–amine	Epoxy–amine
Corrosion protection[b] (h)	>240	>240	>300	>1 000	>1 500

[a] Based on solids as delivered.
[b] ASTMB117, Salt spray test on phosphated steel.

TABLE 6
Bath and Paint Properties of CED for the Automotive
Industry

Solids content[a] (%)	15–25
Specific conductivity at 20°C (mS/cm)	0·8–1·5
pH	4·0–7·0
Ash content (%)	1–5
Voltage (V)	100–450
Bath temperature (°C)	25–35
Time (s)	120–180

[a] 2h/130°C.

automotive industry in 1965 there was a rapid changeover from the anodic process to the cathodic process. Advantages have been seen in the improvement in throwing power, reduced energy consumption, lower film thickness and better corrosion protection (Table 6). Many attempts have been made to change automotive paint technology by developing new paint systems. In one such system, the process was reversed; a primer as powder or a solvent-borne paint was first applied, baked and then the mostly anodic electrocoat process was carried out. This process should give better corrosion protection inside the articles. Howver, because of the excellent throwing power of the cathodic electrocoatings this process is not important today. A similar fate befell a related process, where firstly a cathodic electrocoat, in which fine powdered resins were dispersed, was applied with film thicknesses of more than 0·05 mm.[140] Then the normal electrocoat material was used. This technology requires high equipment investment but does not yield sufficient improvement in quality to warrant the expenditure.

At the beginning of the 1980s high-build cathodic electrocoatings of 0·035 mm thickness entered the scene. These electrocoatings have been successful, especially in the United States, because of existing stringent environmental requirements which are easily met by using such coatings and then reducing the amount of sprayed coating.

The claims for the new generation of electrocoats are:

(a)   better adhesion to the different new materials used for body construction;
(b)   better flexibility for improved stone chip resistance.

In 1985 roughly 40% of the electrocoatings were anodic and 60% cathodic; between them, they took a 2·5% share of total paint business in Europe.[141]

REFERENCES

1. GOULD, R. F. (Ed.), *Electrodeposition of Coatings*, *Amer. Chem. Soc. Adv. Chem. Ser.*, 1973, **119**.
2. MADEN, W., *Handbook of Electropainting Technology*, Electrochemical Publ. Ltd, Ayr, UK, 1978.
3. DAVEY, W. P., US Patent 1 294 627, 1919.
4. CROSSE AND BLACKWELL LTD, British Patent 453 810, 1937.
5. SUMMER, G. C., *Trans. Faraday Soc.*, 1940, **36**, 272.
6. BURNSIDE, G. L. and BREWER, G. E. F., *J. Paint Technol.*, 1966, **38**, 96.
7. BREWER, G. E. F., *J. Paint Technol.*, 1973, **45**, 37.
8. LOOP, F. M., *I-Lack*, 1979, **47**, 253.
9. ARLT, K., STREITBERGER, H. J. and JOSTEN, K. H., *Farbe u. Lack*, 1984, **90**, 162.
10. BECK, F., *Prog. Org. Coatings*, 1976, **4**, 1.
11. PIERCE, P. E., *J. Coatings Technol.*, 1981, **53**, 52.
12. EL-ASSER, M. S., VANDERHOFF, J. W., HAMAYUM, A., HO, C. C. and ABDEL-BARY, M. F., *J. Coatings Technol.*, 1984, **56**, 37.
13. KRYLOVA, A. and ZUBOV, P. I., *Prog. Org. Coatings*, 1984, **12**, 129.
14. FINN, R. S. and MELL, C. C., *J. Oil Colour Chemists Assoc.*, 1964, **47**, 219.
15. TAWN, A. R. H. and BERY, J. R., *J. Oil Colour Chemists Assoc.*, 1965, **48**, 790.
16. BECK, F., *Farbe u. Lack*, 1966, **72**, 218.
17. SPOOR, H. and SCHENCK, H. U., *Farbe u. Lack*, 1982, **88**, 94.
18. SCHENCK, H. U., SPOOR, H. and MARX, M., *Prog. Org. Coatings*, 1979, **7**, 1.
19. PPG, US Patent 4 081 134, 1978.
20. PPG, US Patent 3 962 165, 1976.
21. PPG, US Patent 3 937 679, 1976.
22. PPG, US Patent 3 894 922, 1975.
23. BASF, German Patent 1 546 848, 1965.
24. BASF, German Patent 1 546 854, 1965.
25. BASF, German Patent Appl. 2 002 756, 1970.
26. PPG, US Patent 3 853 803, 1973.
27. ASAHI, Japanese Patent Appl. 56/93728, 1979.
28. ASAHI, Japanese Patent Appl. 57/123216, 1981.
29. ASAHI, Japanese Patent Appl. 58/160311, 1982.
30. BASF, German Patent Appl. 2 924 756, 1979.
31. BASF, German Patent Appl. 3 017 603, 1980.
32. BASF, German Patent Appl. 3 133 770, 1981.
33. PPG, US Patent 3 883 483, 1973.
34. BASF, German Patent Appl. 1 546 840, 1965.
35. SHINTO PAINT, German Patent Appl. 2 324 019, 1972.
36. SHINTO PAINT, German Patent Appl. 2 325 177, 1972.
37. BASF, German Patent Appl. 2 914 297, 1979.
38. BASF, German Patent Appl. 2 914 331, 1979.
39. CELANESE, US Patent 3 716 402, 1971.
40. PPG, US Patent 3 992 253, 1975.
41. PPG, US Patent 3 799 854, 1974.
42. CELANESE, US Patent 4 339 369, 1981.

43. PPG, US Patent 3 853 803, 1974.
44. WISMER, M. L., PIERCE, P. E., BOSSO, J. F., CHRISTENSON, R. M., JERABEK, R. D. and ZWACK, R. R., *J. Coatings Technol.*, 1982, **54**, 35.
45. PPG, US Patent 3 619 398, 1971.
46. PPG, US Patent 3 839 252, 1974.
47. CELANESE, US Patent 4 139 510, 1979.
48. DESOTO, US Patent 4 533 682, 1983.
49. DESOTO, US Patent 4 480 083, 1983.
50. PPG, US Patent 4 017 438, 1971.
51. PPG, US Patent 3 947 339, 1976.
52. BASF, European Patent 134 983, 1983.
53. BASF, European Patent 184 152, 1984.
54. BAYER, German Patent Appl. 2 528 212, 1975.
55. FORD, European Patent 83 232, 1981.
56. VIANOVA, European Patent 28 401, 1979.
57. VIANOVA, European Patent 28 402, 1979.
58. KANSAI PAINT, German Patent Appl. 3 109 282, 1980.
59. BASF, German Patent Appl. 2 320 301, 1973.
60. BASF, German Patent Appl. 2 419 179, 1975.
61. BASF, German Patent Appl. 2 541 801, 1975.
62. BASF, German Patent Appl. 2 711 385, 1977.
63. DESOTO, US Patent 4 341 681, 1981.
64. KEMPTER, F. E., GULBINS, E. and CAESAR, F., *XIV Fatipec Congress, Budapest, 1978*, pp. 297–304.
65. BASF, European Patent 123 140, 1983.
66. NIPPON OIL, German Patent Appl. 2 616 591, 1975.
67. ACC, US Patent 3 984 382, 1975.
68. BASF, German Patent Appl. 3 244 990, 1982.
69. BASF F & F AG, European Patent 59 885, 1981.
70. PPG, German Patent 2 701 002, 1976.
71. PPG, US Patent 4 468 307, 1983.
72. PPG, US Patent 4 148 772, 1977.
73. BASF F & F AG, German Patent Appl. 3 329 693, 1983.
74. BASF F & F AG, European Patent 123 879, 1983.
75. KANSAI, US Patent 4 036 795, 1974.
76. FORD, US Patent 4 536 558, 1983.
77, BASF, German Patent Appl. 3 146 640, 1981.
78. BASF F & F AG, European Patent 154 775, 1984.
79. ACC, US Patent 4 484 994, 1983.
80. KANSAI, German Patent Appl. 2 926 001, 1978.
81. KANSAI, German Patent Appl. 2 934 172, 1978.
82. FORD, US Patent 4 486 571, 1983.
83. BASF, German Patent Appl. 2 755 906, 1977.
84. KANSAI, US Patent 4 190 564, 1976.
85. CELANESE, US Patent 4 139 510, 1978.
86. CELANESE, US Patent 4 190 564, 1976.
87. CELANESE, US Patent 4 137 140, 1978.
88. CELANESE, US Patent 4 134 864, 1978.

89. BASF F & F AG, European Patent 154 724, 1984.
90. BASF, German Patent Appl. 2 755 908, 1977.
91. AKZO, German Patent Appl. 3 014 733, 1980.
92. DULUX, European Patent 69 582, 1981.
93. PPG, European Patent 70 550, 1981.
94. M & T, German Patent 2 057 799, 1969.
95. BASF, European Patent 125 438, 1983.
96. WICKS, Z. W., Prog. Org. Coatings, 1985, 3, 73.
97. HERBERTS, European Patent 4 090, 1978.
98. SHELL, European Patent 12 463, 1978.
99. SHELL, European Patent 40 867, 1980.
100. SHELL, European Patent 79 629, 1981.
101. HERBERTS, European Patent 135 909, 1983.
102. VIANOVA, European Patent 131 127, 1984.
103. BASF, European Patent 167 029, 1984.
104. FORD, European Patent 124 187, 1983.
105. FORD, European Patent 147 049, 1983.
106. BASF, European Patent 123 880, 1983.
107. BASF, European Patent 179 273, 1984.
108. BASF, German Patent 1 930 949, 1969.
109. FORD, US Patent 4 476 259, 1983.
110. PPG, German Patent 2 363 074, 1972.
111. SHELL OIL, US Patent 3 719 626, 1973.
112. VIANOVA, German Patent 2 836 830, 1977.
113. FORD, European Patent 114 090, 1983.
114. PIERCE, P. E. and COWAN, C. E., J. Paint Technol., 1972, 44, 61.
115. DEBYE, P., Ann. NY Acad. Sci., 1949, 51, 575.
116. PIERCE, P. E., KOVAC, J. and HAGGENBOTHAM, C., Ind. Eng. Chem. Prod. Res. Dev., 1978, 17, 317.
117. ANDERSON, P. G., MURPHY, E. J. and TUCCI III, J., J. Coatings Technol., 1978, 50, 38.
118. BECK, F., in Comprehensive Treatise of Electrochemistry, Vol. 2, eds J. O'M. Bockris, B. E. Conway, E. Yeager and R. E. White, Plenum Publishing Corp., 1981, New York, p. 537.
119. SAATWEBER, D. and VOLLMERT, B., Angew. Makromol. Chem., 1969, 8, 1.
120. GIBOZ, J. P. and LAHAYE, J., Paint Technol., 1971, 43, 79.
121. COOKE, B. A., Paint Technol., 1970, 34, 12.
122. BECK, F., Bunsenges Ber. Phys. Chem., 1968, 72, 445.
122. HEIMANN, U., BASF L & F AG, unpublished experiments.
124. HÖNIG, J., J. Oil Colour Chemists Assoc., 1977, 60, 284.
125. GILCHRIST, A. E. and SHUSTER, D. O., 161st meeting, Amer. Chem. Soc., 1971, 346.
126. FURUNO, N. and OHYABU, Y., Prog. Org. Coatings, 1977, 5, 201.
127. WARNECKE, H. J. and BAUMGÄRTNER, O., Z. Ind. Fert., 1981, 71, 353.
128. HEIMANN, U., DIRKING, T. and STREITBERGER, H. J., Surtec '85, 1985, DE Verlag GmbH, Berlin–Offenbach, p. 395.
129. LEBRAS, R. R. and CHRISTENSON, R. M., J. Paint Technol., 1972, 44, 63.
130. PPG, US Patent 3 749 657, 1969.

70   H.-J. STREITBERGER AND R. P. OSTERLOH

131. SPRINGER, W. S., STROSBERG, G. G. and ANDERSON, J. E., *Org. Coat. and Plast. Chem.*, *182nd meeting, Amer. Chem. Soc., New York*, 1981, **45**, 104.
132. ICI, British Patent 1 109 979, 1964.
133. PPG, European Patent 0 156 341, 1985.
134. INOUE, A. and LEGATSKI, K., *Proc. Electrocoat '86*, 1986, p. 14.
135. ZIEGLER, B., *I-Lack*, 1971, **39**, 331.
136. KITAYAMA, M., AZAMI, T., MIURA, N. and OGASAWARA, T., *Trans.*, 1984, **24**, 74.
137. HANSEN, H., *I-Lack*, 1982, **50**, 19.
138. RAUSCH, W., *I-Lack*, 1981, **49**, 413.
139. COOK, B. A., *Proc. 17th Internat. Conf. Org. Coatings Sci. and Technol., Athens*, 1983.
140. ZIMMERMANN, R., *Defazet*, 1979, **33**, 142.
141. Based on *Fatipec Yearbook*, 1986.

CHAPTER 3

Acoustic Emission Testing of Coatings

REES D. RAWLINGS

*Department of Materials, Royal School of Mines,
London, UK*

1. INTRODUCTION

A component's performance in service may often be improved by isolating the main structural material from the environment by use of a protective coating. The coating is generally present to prevent chemical reactions with the environment, but in some cases other characteristics, such as the wear behaviour or the thermal properties, are paramount. The protective coating may also enhance the aesthetic qualities of the component. For simplicity, coatings may be considered under two major groupings, namely the organic coatings and the inorganic coatings. Organic coatings, most of which are known as paints, are generally composed of pigments dispersed in organic binders. The pigments are present to increase the hiding properties, to obtain different colours and sometimes to improve corrosion resistance. The most widely encountered surface coatings in the second grouping are electrodeposited and hot-dipped metallic systems and thermally sprayed coatings, but also included are the more unusual coatings, e.g. diffusion coatings and those produced by laser treatments.

A new coating undergoes a variety of tests before it is put on the market. These fall into two basic categories: firstly, tests designed to assess the production of the coating; and secondly, tests aimed at determining the performance properties. In recent years acoustic emission (AE) has been successfully employed in both categories of investigation and this chapter reviews this work.

1.1. Principles of Acoustic Emission (AE)

When a dynamic process occurs in or on the surface of a material some of the energy which is rapidly released generates elastic stress waves, i.e. vibrations within the material. These stress waves propagate through the material as longitudinal and shear waves and eventually reach the surface, so producing small temporary surface displacements. In extreme cases, for example in the well-known cracking of ice and twinning of tin (tin-cry), the stress waves may be of high amplitude and low frequency and consequently audible. Usually, however, the stress waves are of low amplitude and high frequency outside the audible range of the human ear, and sensitive transducers are required to detect and amplify the very small surface displacements associated with the waves. Many workers have pointed out that, as the 'noise' does not generally fall within the audible range, acoustic emission (AE) is a misnomer and the alternative term 'stress wave emission' (SWE) is a more accurate description of the phenomenon. Both terms are employed, but acoustic emission (AE) is the more widely used.

The most commonly used transducers are piezo-electric crystals which convert the surface displacements into electrical signals. Broad-band transducers are available, but they are more expensive and their sensitivity is less than that of resonant transducers; these latter are capable of detecting displacements equal to or greater than 10^{-14} m and hence are the norm except in frequency analysis work. A high proportion of an acoustic wave is reflected at a solid/air interface; therefore a coupling medium is required to improve the efficiency of transmission of the waveform into the receiving transducer. Good coupling may be achieved by highly viscous, sticky materials such as certain greases, resins and glues. The couplant should be thin—as a thick layer can lead to reflections within the couplant—and of an even thickness. It is necessary to ensure that a couplant will not degrade or change its acoustic properties with time if long-term tests are to be conducted. Further information on couplants may be found in the literature (see ref. 1).

The electrical signal from the transducer is subsequently amplified and the signal resulting from a single surface displacement will be similar to those shown in Fig. 1. In the idealised case (Fig. 2) the voltage V versus time t relationship for such a signal approximates to a decaying sinusoid:

$$V = V_\mathrm{P} \sin 2\pi f t \exp(-t/\tau) \tag{1}$$

where f is the resonant frequency of the transducer, τ is the decay time and V_P is the peak voltage or amplitude. The problem is then one of quantifying the numerous signals which may be detected during a test. A number of

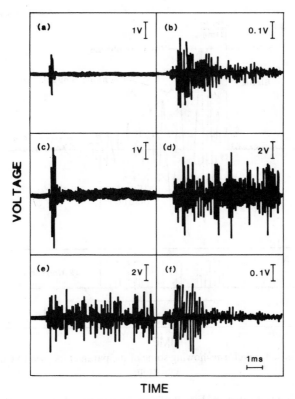

FIG. 1. Examples of AE signals.[2]

signal analysis techniques have been proposed, but only those frequently encountered will be described.

The simplest method to obtain an indication of AE activity is to count the number of amplified pulses which exceed an arbitrary threshold voltage V_t. This is ring-down counting[3] and the signal in Fig. 2(a) would correspond to 12 ring-down counts [Fig. 2(b)]. If the signal approximates to the decaying sinusoid equation, then the number of ring-down counts N_R depends on the peak voltage and is given by:

$$N_R = f\tau \ln(V_P/V_t) \qquad (2)$$

The threshold voltage is, for convenience, usually set at 1 V and the total system amplification, or gain, is often around 90 dB. A 90 dB gain means that the signal is amplified about 32 000 times and hence the 1 V threshold corresponds to a 30 μV signal from the transducer.

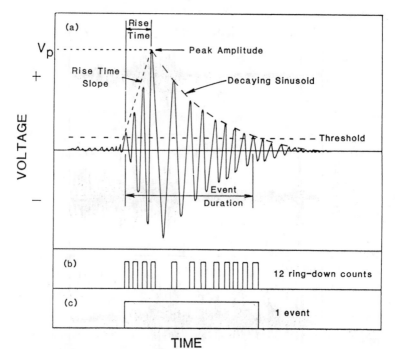

FIG. 2.   Idealised waveform showing some of the parameters used to characterise
emissions.

As the signal shown in Fig. 2(a) can be produced by a single surface displacement, which in turn is often assumed to be the consequence of a single source event inside the material, recording a count of unity is sometimes more convenient than the multiple count obtained by ring-down counting. This objective may be easily achieved electronically by the judicious choice of dead times and this mode of analysis is known as event counting [Fig. 2(c)].

The peak voltage is a function of the AE energy E and for a resonant transducer with narrow-band instrumentation, the appropriate relationship is:[4]

$$E = gV_P^2 \qquad (3)$$

where g is a constant. The acoustic energy is related to the energy of the source event although the exact partition function is generally not known. Furthermore, when narrow-band energy measurements are carried out, only a part of the frequency spectrum is monitored and the distribution of

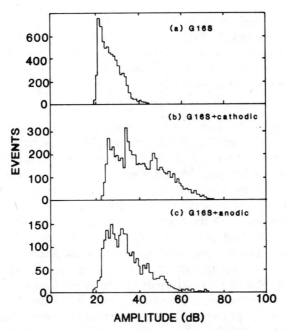

FIG. 3. Typical amplitude distributions:[5] (a) G16S phosphate; (b) G16S phosphate plus cathodic electrocoat; (c) G16S phosphate plus anodic electrocoat.

energy may vary with frequency. Thus, the energy measured over a given
narrow frequency band will not be a known constant fraction of either the
total AE energy or the total energy of the source events.

Some equipment has the facility to record and store the number of ring-
down counts produced by each event, and because of the relationships
between V_P and E and N_R [eqns (2) and (3)], this parameter is an indication
of the AE energy of the events. However, more comprehensive information
on the acoustic energy emitted during a test may be obtained from
histograms of the number of events against V_P or against E; these
histograms are known as amplitude and energy distributions respectively
and typical examples of the former are shown in Fig. 3. Note that as the
peak voltages recorded during a test vary by several orders of magnitude it
is common practice to use a logarithmic, or decibel, scale for the amplitude.
Holt and Evans[6] have demonstrated that, despite complications due to
multiple reflections, an experimental amplitude distribution is a good
representation of the true distribution of stress wave amplitudes in a
specimen.

Several models have been proposed for describing amplitude distributions; these include Rayleigh, extreme value function, Lorentzian, lognormal and power law; see refs 6–11. Of these the most widely employed is the power law which, if only one type of source event is occurring over the monitoring period, is given by:

$$n_a = (V_a/V_0)^{-b} \qquad (4)$$

where n_a is the fraction of the emission population whose peak voltage exceeds V_a, V_0 is the lowest detectable amplitude and the exponent b is a constant which characterises the distribution. The power law has been applied to seismological data for many years[12,13] but was first recognised as being appropriate to modern AE analysis of materials by Pollock.[10,11] An important feature of this model is that the value of b does not change if the amplitudes of all the events are reduced by the same factor. Thus attenuation in a sample, or structure, should not affect the b-value.

In certain situations the source events are taking place so rapidly that the acoustic signals overlap and, in extreme cases, an almost continuous signal may result. Significant overlapping of signals can lead to errors with the previously described counting modes, particularly if the signals are of low energy (low V_p). In these circumstances the measurement of the root mean square voltage (RMS) or the true mean square voltage (TMS) is advisable; however, the voltage must be changing relatively slowly with time, because the response time of most RMS/TMS meters is of the order of 100 ms. The RMS of a time-dependent voltage $V_{(t)}$ over the time interval 0 to T is given by:

$$RMS = \left\{ 1/T \int_0^T V_{(t)}^2 \, dt \right\}^{1/2} \qquad (5)$$

The TMS is simply the square of the RMS and, as the acoustic energy E is proportional to V^2, it is apparent that:

$$E \propto \int_0^T V_{(t)}^2 \, dt = TMS \qquad (6)$$

At present, ring-down and event counting and amplitude distribution are the most often encountered techniques of analysis. Nevertheless, other characteristics of the signal are occasionally used to analyse AE, such as the rise time, rise time slope, event duration (also called the pulse width), energy distribution and spectral (frequency) analysis. The latter technique is worthy of further mention as it may be more widely employed in the future, because of the advent of the procedure of fast Fourier transform associated with the advances being made in computers and the increased sensitivity of broad-band transducers. In spectral analysis the frequencies present and the amount of signal at each frequency are determined. With this information, calculation of the power spectral density is possible as the energy contained in the signal between frequencies f and $f + \Delta f$ is proportional to the area under the frequency distribution graph between these two frequencies.

It must be emphasised that the elastic waves produced by a source event undergo significant modification before and during analysis. Firstly, the waves are attenuated within the material, then on reaching the surface they are reflected, which may even result in the setting up of standing waves. Surface waves may also be produced. The signal is then further altered by the coupling to the transducer, the frequency response of the transducer and the characteristics of the monitoring system. In addition, care must be taken in interpreting the magnitude, and hence the significance, of the various AE parameters associated with the modified signals. An essential aid to the analysis is the waveform as recorded by a transient recorder or equivalent equipment. The importance of the waveform and some of the pitfalls in assessing AE have been discussed by Hamstrad[2] and will be illustrated by taking two examples from the waveforms shown in Fig. 1: (i) only a small change in threshold level, or the amplification, could result in differences of an order of magnitude in the measured duration of the pulse in Fig. 1(c); and (ii) depending on whether the threshold was positive or negative, the measured rise time for the pulse in Fig. 1(e) could vary by as much as a factor of four.

The present state of technology and theory is such that AE alone cannot unambiguously identify source mechanisms on the atomic scale. Generally one has to resort to complementary experiments, e.g. microscopy, to assist in the identification. On the other hand, AE is a convenient technique for detecting the onset of a process, such as cracking of a coating, the extent of that process and any changes that occur as a result of varying the test conditions (e.g. environment, temperature, composition of the coating), as will be illustrated in the following discussion. A book and some articles on AE which have not been referred to in this section, but which may be of interest to the reader, are given in refs 14–19.

2. ORGANIC COATINGS

The author is not aware of any publication on AE and the production of organic coatings, although work is in progress on monitoring the drying of paint finishes.[20] In contrast, the performance of paint finishes has received considerable attention and many aspects of the coating, e.g. composition, stoving temperature, thickness, and the effect of liquid environments, ageing and weathering have been investigated.

2.1. Direct Monitoring of Corrosion

Acoustic emission has been used extensively to monitor corrosion reactions[21-25] and a logical extension of this work is to determine the failure of a paint finish by detecting the corrosion of the substrate. Clearly this approach can only be considered to be viable if detection of emissions is possible when a small area of substrate is exposed to the environment. Scott[26] simulated paint failures by removing 2 mm squares of a lacquer coating from an aluminium sheet. On placing a microlitre drop of 0·16M sodium hydroxide solution on the bare metal substrate, signals were detected by various transducer–amplifier combinations, e.g. transducers with resonant frequencies 50, 100 and 220 kHz were used. This experiment demonstrated the sensitivity of the AE technique and showed that the resonant frequency of the transducer was not critical. Similar experiments have been carried out by Rettig and Felsen[24] who scribed lines through to the substrate on epoxy- and nylon-coated aluminium. Again a small drop of corrodent was sufficient to initiate AE. Filiform corrosion was produced with the nylon coating and this resulted in bursts of acoustic activity which, it was suggested, might be partially attributed to pressure effects on the organic coating.

Although these experiments have shown that direct monitoring of substrate corrosion is possible in certain circumstances, the technique is unlikely to be applicable to all systems. Most of the emissions observed during corrosion are associated with gas bubbles and, therefore, corrosion reactions which involve no gaseous products are relatively quiet. Unfortunately this means that the important corrosion reaction of steel in water and chloride solutions is difficult to monitor. Thus, in the foreseeable future, direct AE monitoring of substrates will be restricted to liquid/substrate combinations which result in bubble formation, such as steels in acids and aluminium and its alloys in chloride solutions.

2.2. Mechanical Testing and Effect of Coating System Variables

Because of the limitations of the direct monitoring technique, work has concentrated on assessing the quality of a coating by recording the emissions from the coating plus substrate during a tensile test. Possible AE source mechanisms associated with a paint finish and the underlying

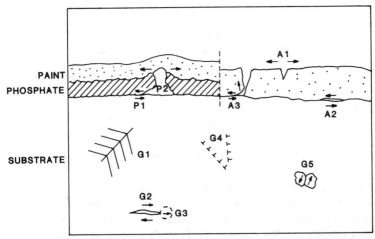

FIG. 4. Possible sources of AE in a complete paint system, including the substrate.[27] A1, crack formation and growth in paint; A2, micro-delamination of the paint; A3, delamination; P1, delamination of the phosphate layer; P2, cracking of the phosphate layer; G1, twinning; G2, crack face rubbing; G3, crack growth; G4, slip; G5, inclusion decohesion or fracture.

metallic substrate are presented in Fig. 4.[27] As will be seen in the following sections, a wide range of paint finishes have been investigated on mild steel and aluminium. Both of these metals are ductile and give few emissions when deforming compared with most coatings. To the author's knowledge mechanical testing of paint finishes on wood has not been reported in the literature, probably because of the greater number and variability of the emissions from wood.

2.2.1. Effect of substrate

The substrate material and the pretreatment it has received can affect the performance of a paint system. There is a dearth of information on the effect of substrate material as most studies have been carried out on steel substrates. However, a comparison of the behaviour of a series of epoxy resins on aluminium and steel revealed only slight differences in acoustic response; the accumulative ring-down count versus strain plots were of similar form for both substrates but the level of the curves showed a greater variation with the chemistry of the epoxy on aluminium than on steel.[28,29]

More significant differences in the acoustic characteristics have been found with the pretreatment received by steel substrates. A common pretreatment is phosphating, which is the spraying or dipping of steel in a

TABLE 1
Characteristics of the Phosphate Coats[5]

Code	Coating weight ($g\,m^{-2}$)	Hopeite needles	
		Density	Length (μm)
G902	1·8	Low	20–30
G2500	1·9	High	30–40
G2000	2·0	Low	20
G16	2·3	High	40
G16S	2·6	High	20–40

solution of phosphoric acid, a metal phosphate and an accelerator to form a conversion coating. Phosphating is thought to inhibit the spread of underfilm corrosion. It is also often quoted as improving adhesion by a keying mechanism, but some of the following AE studies cast a doubt on this particular attribute of phosphating.

Five zinc phosphates (Table 1), differing in microstructure and coating weight, have been investigated by Rooum.[5] On tensile testing, all the phosphated steels were noisier than degreased steel, although the phosphate G902 was only marginally so; typical event count versus strain plots and amplitude distributions from this study are given in Figs 3 and 5. It would appear that the acoustic emission is determined by both the coating weight and the microstructure. For constant microstructure, e.g. phosphates G902 and G2000 with a low density of small (20–30 μm) hopeite crystals, or phosphates G2500, G16 and G16S with a high density of larger (20–40 μm) hopeite crystals, more emissions were observed the greater the coating weight. Scanning electron microscopy indicated that the sources of emission were the cracking and loss of adhesion of the hopeite needles and cracking of the underlying phosphophyllite platelets. These sources operated to varying extents in the five phosphates but, as can be seen from the amplitude distribution of Fig. 3(a), which is typical of all the phosphates, they are relatively low-energy sources. Results from other workers[30-32] are in general agreement with those of Rooum in that the number of emissions obtained from zinc phosphated steels was greater than that from degreased steel and increased with coating weight. In contrast, it has been reported that an iron phosphated steel can give fewer emissions than a degreased steel.[27]

Data are available on the effect of phosphating on the performance of electrocoats (cathodic and anodic) and epoxy resins (liquid and powder). In the case of the electrocoats, the cathodic was more sensitive to the type of phosphate than the anodic.[5] The former systems were noisier than either

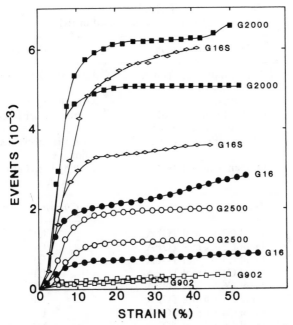

FIG. 5. Accumulative event count versus strain curves for various zinc phosphate coatings (see Table 1) on steel.[5]

the phosphated steel or the anodic systems. The electrocoats produced considerable acoustic activity in the early stages of a test but, unlike the behaviour for the phosphates, this was followed by another burst of emissions at higher strains. With an anodic coating these later emissions were at final failure; in contrast, for the cathodic systems they commenced at strains in the range 16–28%, with the onset at a lower strain the greater the phosphate coating weight. The amplitude distributions for the electrocoated systems extended to much higher amplitudes than for the phosphated steels, indicating an additional failure mechanism associated with the electrocoat (Fig. 3).

Acoustic emission studies have been made on a series of unpigmented epoxy resins, in which the chemical structure had been modified by reaction with various reagents (Table 2); the resins were applied to degreased and phosphated steel.[28,33] When coated on phosphated steel the more rigid epoxies, i.e. those that had been extended with aromatic substituents, exhibited a definite increase in the number of ring-down counts in the strain range 5–9% compared with that observed with the degreased substrate. This acoustic activity was attributed to loss of adhesion between the

TABLE 2
The Unpigmented Epoxy Coatings used in Reference 28

Series C	Chemical description	Special features
C1	Tall-oil fatty acid (TOFA)	Aliphatic ester linkages
C2	p-t-Butylbenzoic acid	
	(PTBBA)	Aromatic ester
C3	p-t-Butylphenol (PTBP)	Aromatic ether linkage
C4	Diphenylolpropane (DPP)	
C5	p-Phenylphenol (PPP)	Aromatic substituted ether linkage
C6	Diethanolamine (DEA)	ca 10 hydroxyl groups per molecule

Series D	Derived from reaction of	Cross-linked with
D1	1 mol C2 + 2 mol TOFA	UF
D2	1 mol C2 + 4 mol TOFA	UF
D3	1 mol C3 + 2 mol TOFA	UF
D4	1 mol C4 + 4 mol TOFA	UF
D5	1 mol C6 + 2 mol TOFA	UF
D6	1 mol C6 + 4 mol TOFA	UF
D7	1 mol C6 + 6 mol TOFA	UF
D8	1 mol C6 + 8 mol TOFA	MF
D9	1 mol C6 + 6 mol TOFA	MF
D10	1 mol C4 + 4 mol TOFA	MF
D11	1 mol C2 + 4 mol TOFA	MF

coating and the phosphated substrate. The adverse effect of phosphating on adhesion was confirmed using a hesiometer-type adhesion test, which measured the force to separate the paint finish from the substrate to be 2·20 and 0·79 kg for the degreased and phosphated conditions respectively.[29]

Hansmann and Moslé[27,34] have shown that differences in adhesion between an epoxy powder coating and zinc and iron phosphate pretreatments can be detected by AE. The poor adhesion associated with the iron phosphate resulted in fewer emissions than were recorded for the zinc phosphate. Furthermore, AE analysis revealed iron phosphating also to be an unsatisfactory pretreatment for a liquid epoxy coating, while zinc phosphating was only slightly, if at all, inferior to degreasing as a pretreatment.

2.2.2. Effect of production variables
Moslé and Wellenkötter[30,32] were the first to report that modifications to coatings brought about by variations in stoving temperature may be

conveniently studied using AE. A later investigation by Hansmann and Moslé found the response of epoxy powder and liquid coatings to stoving temperature to be different.[27] Stoving temperatures in the range 160–200°C inclusive had little effect on the emissions from liquid coatings, whereas the AE data indicated better adhesion at the higher stoving temperatures for the powder coatings. These results demonstrate the sensitivity of the AE technique as these workers reported that the 'usual paint test methods' were unable to detect any differences in the adhesive strength either between the two paint systems or with stoving temperature. In addition to ring-down counting a parameter, termed the C-value, which has units of voltage and is a measure of the overall acoustic energy, has been employed in a study of the effect of stoving temperature on polyester coatings.[35] In simple terms the C-value depends on the extent of the damage during the mechanical test, e.g. area of delamination, and on the magnitude of the relevant forces, e.g. cohesive or adhesive. With a few exceptions—see Section 2.3.1—the lower the C-value the better the condition of the paint system under test. For a polyester coating on degreased steel the C-values were 4·0, 4·7 and 2·0 V for stoving temperatures of 180, 200 and 220°C respectively.

Under normal conditions of application the thickness of a paint coating will vary considerably and consequently it is important to know to what degree the AE is modified by such variations. Work on unpigmented acrylic-based varnishes of thickness in the range 20–80 μm inclusive found no major differences in strain dependence of the ring-down count; for the thick 80 μm sample, the count rate was just slightly higher and the onset of macroscopic cracking was at a somewhat lower strain.[31,32,36] This result suggests that normal fluctuations in thickness will give negligible scatter in AE data; however, as pointed out by Hansmann and Moslé,[37] it is preferable to compare the results from coatings of equal or similar thicknesses.

Many paint finishes which are placed in service are multilayer systems consisting of, for example, a primer, undercoat and topcoat. By studying individual coats as well as different combinations of component coats that make up the full systems, useful information can be gained on the interactions of the component coats and this can lead to a better understanding of the performance of the full paint finish.

Moslé and his coworkers[31,32,35] have shown that the AE count rate is affected by the presence of an undercoat and that the AE characteristics alter when different undercoats are used with the same top-coat. In the case of an unspecified paint system consisting of a primer, first undercoat, second undercoat and topcoat, a well-adhering second undercoat resulted

TABLE 3

Peak Positions and Intensities of the Peaks in the Amplitude Distributions at 10% Strain and Failure for Full System and Subsystems of an Automotive Finish[39]

Specimen	Measurement point	Intensity[a]				
		Peak position 22±1dB	Peak position 26±1dB	Peak position 35±1dB	Peak position 47±2dB	Peak position >60dB[b]
Steel/phosphate	10% strain	2	1			
	Failure					†
Steel/phosphate/primer	10% strain	2	1			
	Failure			2	3	†
Steel/phosphate/primer/surfacer	10% strain		1			†
	Failure			2	3	†
Steel/phosphate/primer/surfacer/topcoat (full system)	10% strain		1	2		†
	Failure		1	2		†
Failure process		Hopeite cracking	Adhesion loss	Peeling and microcracking	Peeling and microcracking	Microcracking

[a] Key to intensity code: 1, high intensity; 2, medium intensity; 3, low intensity.
[b] There were insufficient events above 60dB for the computer program to fit peaks. However, the dagger (†) indicates the presence of a small number of high-amplitude events.

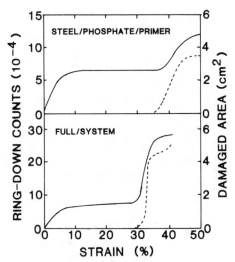

FIG. 6.   Graphs showing the good correlation between the extent of gross damage, given by the broken lines, and the accumulative ring-down count at high strains for an automotive finish.[39]

in fewer emissions than a second undercoat to which adhesion-reducing agents had been deliberately added.[31,32]

An even more complex multilayer system is the automotive finish, which usually consists of a steel substrate with a phosphate layer, an electrocoat primer, a surfacer and a topcoat. The AE from an automotive finish and several subsystems of the finish, complemented with information from other techniques such as scanning electron microscopy, has enabled the identification of a number of failure mechanisms and the strain range over which they occurred.[38,39] In particular, a good correlation was obtained between the accumulative ring-down count and the extent of visible surface damage at high strains (Fig. 6). Also in this work emphasis was placed on amplitude distributions, which varied considerably between the subsystems and the full system (Fig. 7). Analysis of the amplitude distributions in terms of overlapping Lorentzian peaks established that each failure mechanism was associated with emissions of a characteristic amplitude, as summarised in Table 3.

2.2.3.   Effect of binder composition and pigment concentration
Acoustic emission data which compare completely different paint systems in the as-prepared condition have been reported in the literature, e.g. epoxy and acrylic,[30] polyester/melamine, maleinised oil and epoxy,[40] spin-coated

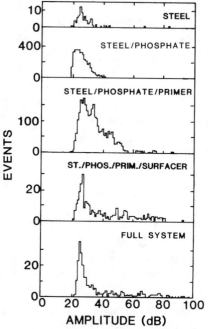

FIG. 7. Typical amplitude distributions at failure for subsystems and the full
system of an automotive finish.[39]

polybutadiene and 0·9 wt fraction epoxy/polyamide.[41] That AE is sensitive
to more subtle variations in composition has been demonstrated by Bahra,
Strivens and Williams-Wynn[28] in their study of unpigmented epoxy
coatings. Two series of modified epoxy resins were used for this study
(Table 2). In series C the terminal glycidyl groups of Epikote resins were
reacted with various reagents to produce different backbone terminal
groups. In the second series (series D), three of the resins (C2, C3 and C6)
were further reacted with tall-oil fatty acid (TOFA) to give resins D1 to
D11. The resins of both series were cross-linked with urea–formaldehyde
(UF), except D8 to D11, for which butylated melamine–formaldehyde
(MF) was used.

The AE results from this investigation are presented in Fig. 8 and it can
be seen that these correlate well with the molecular structure. For instance,
any modifications which led to a decrease in hardness and increase in
flexibility, e.g. extending the backbone with long-chain fatty acid (C1) or
substituting such groups along the backbone (D series), reduced the
number of emissions. The general relationship between the AE and the

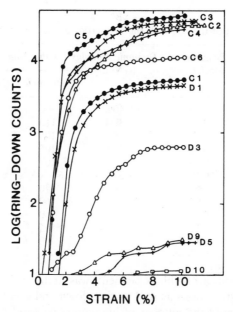

FIG. 8. Accumulative ring-down counts as a function of strain for two series (C and D; see Table 2) of epoxy coatings on aluminium.[28]

mechanical properties of the coatings, as determined by a dynamic method and an indentation technique, is illustrated in Table 4. However, as pointed out by the investigators, the mechanical properties data do not show the discrimination of chemical structure changes revealed by the AE analysis.

The effect of anti-corrosion pigments on the AE from maleinised oil and epoxy resin paints, tested in the as-prepared state, has been shown to be different.[40] The magnitude and strain dependence of the ring-down count was similar for the maleinised oil paint with and without the pigment, whereas for the epoxy binder the presence of the pigment greatly increased the acoustic activity over the complete strain range of the test. No cracking of the pigment-containing epoxy resin coating was observed at low strains to account for the emissions, instead, it was postulated that the sources of emission were micro-adhesive failures at the binder/pigment interface and possibly rubbing between the pigment particles and between the pigment and the substrate.

A more detailed investigation of the role of pigments in determining the AE from a paint finish has been carried out by Miller and Rawlings.[42] An acrylic binder was employed and the effect of concentrations of up to 50%

TABLE 4
Correlation of the AE with the Properties of the Epoxy Coatings[28]

Specimen	Penetration depth (μm)	Dynamic modulus (GN/m²)	Glass transition temp. (°C)	
			Dynamic	Indentation
C5	0·30	40·8	90	75
C3	0·25	39·4	110	75
C2	0·25	39·7	105	81
C4	0·35	39·8	105	>90
C6	0·35		107	74
C1	0·40	37·1	64/105	38
D1	0·65		70	32
D3	1·05		60	32
D9	>6		0	−15
D5	1·15		65	30
D10	1·85		30	20

(AE activity — increasing upward, from C4 down to D10)

titania (TiO$_2$) and carbon black pigments was studied using amplitude and pulse width distributions as well as ring-down and event counting. Both pigments increased the number of high-amplitude and high-pulse-width events, this trend being more marked the greater the pigment content (Fig. 9). The high-energy events were attributed to the formation of microcracks of 20–300 μm in length and macrodamage, i.e. that visible by the eye. The total event and ring-down counts produced the most noticeable difference between the two pigments; for the carbon black the counts showed a monotonic increase with concentration while for the titania there was a maximum in the concentration range 30–40%. Macrodamage was more

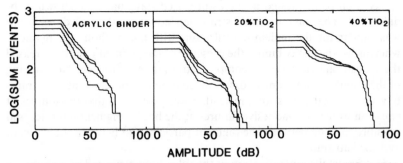

FIG. 9. Amplitude distributions [plotted as log (sum events) against dB] at 10%, 20%, 30% strain and failure for an acrylic binder with and without titania pigment.[42]

extensive with the carbon black and was mainly responsible for the greater acoustic activity for this pigment at high pigment contents.

2.3. Mechanical Testing and Environmental Effects

The properties of a paint finish are known to change with time and at a rate that depends on the environment. Acoustic emission has been employed to determine the degradation of coatings due to immersion in water and salt solution, to exposure to salt spray, to ageing and to weathering (artificial and natural). In all these studies, when other techniques such as micro-indentation, cross-hatch, capacitance and electrochemical impedance were used to monitor the coating, the results obtained were consistent with the AE data. In some cases, AE was found to be the most sensitive technique, capable of detecting subtle changes in the coatings at an earlier stage than was possible by the more conventional methods.

2.3.1. Water immersion

Several workers have found that ring-down counting,[31,38,43,44] and sometimes even RMS monitoring,[38] are capable of detecting the deterioration produced by immersion in water. Some typical results[43] for two full automotive finishes, consisting of a phosphate coat, an electrocoat (either cathodic or anodic), surfacer and topcoat, are presented in Fig. 10. In the dry state the ring-down rate plots had two distinct peaks, that at the higher strains being associated with gross damage (cracking and loss of adhesion) to the paint system (Fig. 6). After water immersion the second

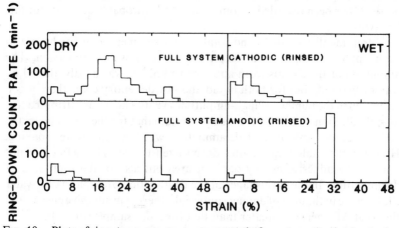

FIG. 10.   Plots of ring-down count rate versus strain for automotive finishes in the as-prepared (dry) state and after 96 h immersion in water at 40°C (wet).[43]

peak occurred at lower strains; this displacement of the acoustic activity to lower strains was slight for the finish with the anodic electrocoat but was so marked for the cathodic system that the high and low strain peaks merged into one. Consistent with the AE data were the observations that gross damage occurred at lower strains after water soaking and that systems with cathodic electrocoats absorbed more water, and retained it for a longer time, than those with anodic electrocoats. Similar work on epoxy powder[31] and acrylic[44] coatings confirmed that the onset of the high-strain AE peak was at progressively smaller strains the longer the specimens were soaked in water. The latter investigation[44] also demonstrated that the type and amount of pigment present affects the response of a coating to water. Both rutile (titania) and zinc phosphate pigments increased the time of soaking required to produce significant acoustic activity at low strains, i.e. the pigments reduced the susceptibility of the acrylic to water induced degradation. Furthermore, as the titania content increased the longer were the immersion times before any deterioration of the finish could be detected by AE. The suggestion was made that this result was a consequence of the longer diffusion paths for the water molecules with increasing pigment content.

In contrast to these examples, water immersion of brittle coatings, such as a weathered paint, can result in the displacement of the high-strain acoustic activity to larger strains and a reduction in the number of counts.[28,29,35,38] The dramatic effect that saturation with water has on the mechanical properties—a lowering of the glass transition temperature by 30–40°C has been recorded in some cases[28,29]—probably accounts for this behaviour.

Amplitude distributions and total AE energy measurements have also been applied to investigations on the effect of water immersion. The amplitude of the signals from automotive finishes was hardly altered by water soaking,[43] but for both liquid and powder epoxy coats there was a reduction in the number of high-amplitude events which was attributed to a loss of adhesion.[35] Clearly the results indicate that for the epoxy coats the AE energy per unit area of delamination was less after water soaking. However, the total energy emitted during a tensile test will be a function not only of the adhesive energy per unit area but also of the total area of delamination. In the early stages of the degradation of a coating by water the increase in the area of damage outweighs the loss in adhesive energy and the total AE energy is greater than that for a dry sample (Fig. 11).[35,37,44] After long soaking times, the fall in the adhesive energy can be so marked that the total AE energy decreases.

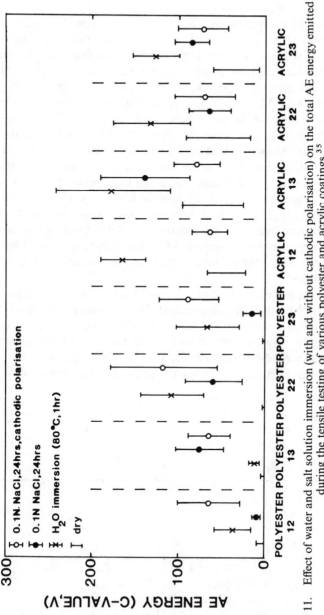

FIG. 11. Effect of water and salt solution immersion (with and without cathodic polarisation) on the total AE energy emitted during the tensile testing of various polyester and acrylic coatings.[35]

2.3.2. Salt solution and spray

Osmotic and electrolytic ion transport through a coating leads to a loss of adhesion and, in the advanced stages of degradation, the formation of corrosion products at the coating/substrate interface. Salt environments have been extensively investigated as they are so frequently encountered in service. The damage on soaking in salt solutions at room temperature has been shown by AE to be progressive and take place over several hours or days depending on the paint system, e.g. for acrylics[44] in 0·5M-sodium chloride, changes were observed within 16 h whereas over 25 days were required before any deterioration of automotive finishes was detected.[42] These studies, and also one on polybutadiene,[41] found that the more marked the damage associated with the immersion, the larger was the number of emissions in the first half of the tensile test. The effect of different pretreatments of the steel substrate on the susceptibility of the polybutadiene coating to salt solution was determined by means of both AE and impedance measurements of the finite pore resistance.[41] Both these techniques ranked the five pretreatments in the same order for resistance to degradation.

Cathodic polarisation of powder and liquid epoxy,[35] polyester,[35] and acrylic[35,44] coatings has been studied. There is general agreement that AE can monitor modifications to coatings brought about by cathodic polarisation, which accelerates the changes produced by simple immersion. The sensitivity of AE was such that the deterioration of an epoxy powder coating was identified after only 100 min of cathodic polarisation, which

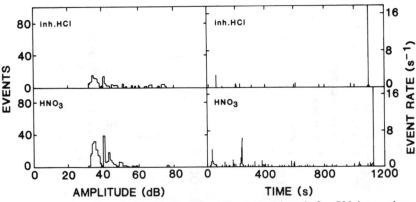

FIG. 12. Acoustic emission from polybutadiene-coated steel after 70 h immersion in 0·5 M-NaCl solution.[41] The pickling in hydrochloric acid produced a system less susceptible to degradation than etching in nitric acid.

compared favourably with periods of 10 h to several days required before the detection of damage by conventional methods.[27]

Whether the signals are of lower or higher amplitude after immersion in salt solutions, with or without cathodic polarisation, would appear to depend on the coating system and the failure mechanisms. This is exemplified by the comparison of results obtained from studies on acrylic and polybutadiene finishes. Undercoat migration from the edges of acrylic specimens can cause cracking after long exposure times, and the energies of the associated AE signals were reported to be higher by a factor of 10–1000 than those of the emissions due to the loss of adhesion without the formation of cracks.[44] In contrast, the substrate pretreatments which resulted in significant degradation of the polybutadiene systems tended to produce a large number of relatively low-amplitude events (Fig. 12), which was attributed to embrittlement of the polybutadiene coating, akin to the behaviour of a brittle epoxy.[41] Hansmann and Moslé[35] measured the total AE energy and found that it increased with respect to the as-prepared coatings in a similar manner to that for water immersion (Fig. 11.).

Salt-spray tests, conforming to ASTM standards,[45] differ from immersion tests in that the spraying is not continuous and consequently the paint can dry out during the periods when the spray is not in operation. Salt-spray testing of the series C epoxy coating listed in Table 2 resulted in a

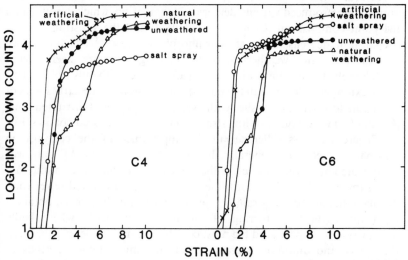

FIG. 13. Effect of exposure to a salt spray and to weathering (natural and artificial) on the AE from two epoxy coatings (C4 and C6; see Table 2) on degreased steel.[33]

decrease in the number of ring-down counts, with respect to the as-prepared coating, for all the finishes except C6 (Fig. 13).[33] The reduction in the emissions for C1 to C5 was attributed to modification of the mechanical properties of the brittle coatings, as previously discussed for water immersion, rather than to any major change in adhesion. Bahra *et al.*[33] suggested that the molecular structure of C6 would enhance the hydrophilicity of the coating, and could have led to greater swelling when wet, and subsequent crazing during the drying periods. The crazing could account for the anomalous increase in AE for this finish.

2.3.3. Ageing and weathering

The important parameters in the weathering of a paint are: (i) time, (ii) temperature, (iii) light, and (iv) environment. The limited amount of information available indicates that the degradation on ageing in a dry atmosphere at room temperature in the absence of light is slight. Automotive finishes, with either an anodic or a cathodic electrocoat, were aged in this manner for five years and then tested dry and after water immersion.[46] The changes in the AE due to this ageing were minor compared with the effects reported in decorative and industrial finishes subjected to a combination of damp, light and heat over a period of time in natural and artificial weathering tests.

Natural exterior weathering of epoxy coatings in an industrial atmosphere produced varying degrees of degradation, which depended on the paint composition (C1 to C6 in Table 2) and the substrate surface preparation (degreased bare steel or phosphated steel).[33] The appearance and the AE, after weathering, of the coatings on bare steel differed considerably; C4, the finish with the best appearance with only isolated rust spots, exhibited a decrease in the number of emissions in the early stages of the tensile test whereas finishes with extensive rusting and disrupted films (C5 and C6) were noisier at low strains than the as-prepared coating (Fig. 13). There was less difference in the appearance of the coatings on phosphated steel, and generally the interpretation of the AE data was complicated by emissions due to adhesive loss. Nevertheless, as for the bare steel substrate, the finishes with the best appearance emitted fewer ring-down counts in the low-strain region. Similar results have been obtained from the natural weathering of automotive finishes[42] where badly degraded finishes produced the most counts. The main difference between the epoxy and automotive coatings was the much longer periods of weathering required before the acoustic response of the automotive finishes changed.

Studies on the AE of coatings which have been subjected to artificial, accelerated weathering[33,42] revealed that badly degraded finishes emitted more noise at low strains than undegraded finishes. Similar observations have been made for coatings which have undergone natural weathering.

3. INORGANIC COATINGS

A wide variety of ceramic and metallic coatings have been investigated using AE although, in general, these have not received as much attention as organic coatings. On the other hand, unlike the case for organic coatings, there have been a number of studies where AE has been employed to monitor the coating process in addition to post-coating assessment.

3.1. Electrodeposition

Corrosion protection of a metal, such as steel, by electrodepositing a layer of a less reactive metal on the surface is standard industrial practice. During electroplating, high residual stresses may be produced in the coating which can lead to cracking of various forms; clearly for the satisfactory protection of the substrate the cracking of the coating must be kept to a minimum. Takano and Ono[47] were the first to demonstrate that acoustic monitoring of the electroplating process can provide information on the quality of the deposit. They recorded the emissions accompanying the electrodeposition of chromium and copper and observed significant differences with changes in the bath conditions. In particular they reported that the acoustic activity during the electroplating of chromium was a consequence of the microcracking due to the residual stresses, a result which has since been confirmed by other workers;[37,48] the rise in the AE counts with increasing crack density in chromium coatings is shown in Fig. 14.

Acoustic emission has also been successfully applied to the monitoring of the electrodeposition of nickel on copper. De Iorio, Langella and Teti[49] investigated the effects of current density, deposition temperature and iron(II) chloride additive, whereas Wellenkötter and Moslé[50] studied the changes resulting from the addition of pyridine chloride brightening agent. Both groups of workers agree that deposition conditions that result in poor quality, i.e. cracked coatings, give a large number of AE events which increase approximately linearly with time over the deposition period. Minor defects may or may not lead to a small number of emissions. For example, no significant AE was produced from high-temperature (40–60°C) deposits of poor surface finish, resulting from the formation of

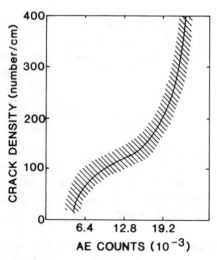

FIG. 14. Acoustic emission activity during the electrodeposition of a 4 µm coat of chromium as a function of crack density.[48]

nickel powder and the presence of burnt surface zones, caused principally by the exhaustion or contamination of the bath. In contrast, a small number of emissions were recorded from rough, matt-finished coatings of variable grain size deposited at 45°C and from visually acceptable coatings produced at ambient temperature from iron(II)-chloride-containing electrolytes. These deposits were not cracked and the emissions were attributed to local plastic deformation in the former and minor damage at the coating/substrate interface in the latter.

The quality of electrodeposits has been assessed by analysing the AE recorded during post-plating tensile test in a similar manner to the work previously described on paint finishes. The metals examined include nickel, copper and zinc, and the effects of the following production variables have been studied: substrate roughness,[32] current density,[51,52] deposit thickness[30,52] and additives.[30,32,51,52] Although amplitude, energy and pulse-width distributions and frequency analysis have been used,[52] most of this work is concerned with the interpretation of differences in count rates. Just one of the metals (zinc) and one of the production variables (thickness) will be discussed in order to illustrate the type of information gained from the AE analysis. As the thickness of the deposit increased the number of counts grew and the higher were the amplitude and energy of the emissions (Fig. 15). This trend in the AE was related to the changes in the failure mode

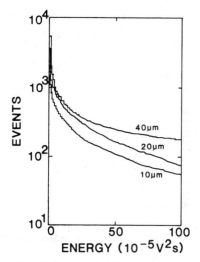

Fig. 15. Effect of coating thickness on the energy distributions [plotted as log (sum events) against energy] from tensile tests on electrodeposited zinc.[52]

with thickness. Thick deposits tended to produce large macrocracks which were visible during testing and were initiated at lower strains as coating thickness increased. Loss of adhesion was also observed in these coatings and this phenomenon, together with the macrocracking, was considered to be responsible for the many emissions of large energy. However, thin deposits were characterised by numerous fine microcracks, which were not visible while the test was in progress, and the low-energy emissions were attributed to these small defects.

3.2. Thermal spraying

In thermal spraying molten particles are propelled by a gas on to a substrate where they solidify to produce a coating which, depending on the coating material and the spraying process, may be from 50 μm to several millimetres thick. The main applications for thermal sprayed coatings are for hardfacing and thermal barriers.

Damage to a coating may occur during the cooling from the spraying temperature and this degradation has been detected for arc and flame spraying by simple counting techniques.[53,54] The AE, and hence by inference the quality of the coatings, was found to depend critically on the spraying parameters such as current intensity and air pressure for arc spraying (Fig. 16) and substrate preheat temperature for flame spraying.

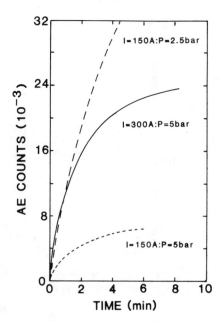

Fig. 16.   Influence of current intensity and gas pressure on the AE recorded during
cooling from arc spraying.[53,54]

The relationship between the spraying parameter and AE was explained by
Steffens and his coworkers[53,54] in terms of the magnitude of the thermal
stresses produced during cooling. For example, as the current intensity
increases the particles become hotter and the temperature differences
between particles and substrates increase. Thus the thermal stresses in the
coating, and hence the extent of the cracking and the number of AE counts,
increased as the current intensity was raised. Similar reasoning accounts for
the observed increase in counts on lowering the substrate temperature or
reducing the air pressure.

A coating may be subjected to thermal stresses not only during
production but also in service; the latter occurs particularly in the case of
thermal-barrier ceramic coatings which often experience large temperature
excursions. Herman and coworkers[55–57] have used ring-down counting to
assess the behaviour during thermal cycling of the following plasma
sprayed ceramics: Al_2O_3, $Al_2O_3 + TiO_2$, TiO_2, CaO-stabilised ZrO_2 and
Y_2O_3-stabilised ZrO_2. This work enabled a comparison of the resistance of
these coatings to single and multiple thermal cycles to be made, and also an
evaluation of the effects of the presence of a NiCrAl bond coat at the

ceramic/substrate interface and of preheating of the substrate. The authors emphasised that great care must be exercised in the interpretation of the AE results because of the complexity of the system. Some of the factors which can affect the type of damage and the AE are the percentage porosity, the presence of metastable phases, the difference in the thermal coefficients of expansion between the coating and the substrate, and the anisotropy of the thermal expansion of the phases present. One of the main findings of this work was that ceramics which exhibit good adhesion to the substrate—or to a bond coat—generate a stable network of microcracks during production or during the first few thermal cycles. These coatings remain effective as thermal barriers even though they contain a network of microcracks and, once the network is established, they are acoustically quiet. In contrast, other coatings fail catastrophically because of unstable crack growth, which is accompanied by many emissions. These two extremes of behaviour are exemplified by stabilised ZrO_2 and Al_2O_3; only about 2000 counts were monitored per cycle for stabilised ZrO_2 with a stable network of microcracks, whereas over 200 000 counts were recorded for each cycle of Al_2O_3. Thus, AE is able to distinguish easily between these two types of coating.

The factors important to the integrity of brittle sprayed coatings have been investigated by monitoring the acoustic activity during mechanical testing. The earliest work[58] recorded the ring-down counts accompanying the indentation of Al_2O_3 and $Al_2O_3 + TiO_2$ coatings by a Brinell Hardness indentor at 500 kg. This preliminary investigation demonstrated that the number of counts varied with the spraying process; for example, oxyacetylene-sprayed Al_2O_3 gave over twice the number of counts as the same material plasma-sprayed. Coating composition also modified the number of counts, $Al_2O_3 + TiO_2$ coatings being considerably quieter than Al_2O_3. Safai et al.[58] suggested that the porosity of the coatings played a significant role in determining the crack morphology and AE, with high porosities producing many emissions. Tensile adhesion tests, in which the coating is pulled from the substrate by means of supports attached to the coating and to the substrate by an epoxy resin, have been used in conjunction with AE monitoring to investigate the effect of substrate preparation and post-production annealing on the performance of a Y_2O_3-stabilised ZrO_2 coating.[59,60] As for the earlier indentation study, ring-down counting was the only analysis technique employed to characterise the emissions. Nevertheless a correlation was found between the adhesive bond strength and the AE. The overall trend observed was that the greater the bond strength, the higher the count rate, with the proviso that a faulty coating can, particularly in the initial stages of the test when the applied load is still low, give an anomalously high count rate.

A more detailed investigation of the stress waves emitted during mechanical testing was carried out by Almond, Moghisi and Reiter.[61] They studied plasma-sprayed molybdenum, self-fusing molybdenum (75% Mo in a low melting point matrix of NiCrBSi) and Al_2O_3 by means of event counts and amplitude distributions obtained from four-point bend tests. Variations from the normal production procedures were chosen to simulate either (i) poor substrate surface preparation by the omission of grit blasting prior to coating or by the use of spent grit, or (ii) inferior spraying technique by running at low power or without cooling. Although both the event counts and the b-values from the amplitude distributions did reveal differences between the various coats, the investigators proposed a statistical analysis of the amplitude distributions which they considered more sensitive and reliable. The statistical method employed was the χ^2

TABLE 5

χ^2 Analysis of the Amplitude Distributions at Different Loads under Four-Point Bending from Seven Samples of Plasma-Sprayed Al_2O_3[61]

Rank	Sample[a]	χ^2	Rank	Sample	χ^2	Rank	Sample	χ^2
	Load 80 kgf			Load 100 kgf			Load 100 kgf	
1	1	0	1	3	11	1	3	12
2	2	0	2	6	23	2	6	37
3	3	0	3	4	27	3	4	47
4	4	0	4	7	61	4	7	60
5	5	0	5	5	78	5	5	118
6	6	0	6	1	226	6	1	290
7	7	0	7	2	759	7	2	1 263
	$F = 180$[b]			$F = 42$			$F = 42$	
	Load 120 kgf			Load 20–120 kgf				
1	3	25	1	6	162			
2	6	39	2	3	209			
3	7	73	3	7	274			
4	4	93	4	4	348			
5	5	164	5	5	502			
6	1	402	6	1	1 258			
7	2	1 819	7	2	4 325			
	$F = 66$			$F = 504$				

[b] F = degrees of freedom.
[a] Samples 2–7 were sprayed under standard conditions; sample 1 was sprayed with dusty grit.

analysis, which enables the scatter in a group of samples to be determined and tests the amplitude distribution of any sample to ascertain whether it lies outside the scatter band or belongs to the same group. The greater the value of χ^2, the larger the difference between the amplitude distributions of the sample and the group. The results for seven samples of Al_2O_3 are presented in Table 5; sample 1 was produced with the substrate surface poorly prepared and samples 2 to 7 were coated under standard conditions. Two samples, numbers 1 and 2, had high χ^2 values and were clearly different from the rest of the group. On examination, sample 2 was found to be cracked and sample 1 had a poorly prepared substrate; hence the conclusion was drawn that the statistical method employed correctly identified the two faulty coatings out of the group of seven.

3.3 Miscellaneous Coatings

In surface treatments the laser can be used either to clad a surface by powder fusion or to alter the surface structure of materials either by a rapid melt/quench or by transformation hardening without melting. The possibility of using AE to monitor and to control laser surface treatments has been examined by Rawlings and Steen.[62] As far as transformation or rapid melt hardening was concerned these workers found that the extraneous emissions associated with the process, e.g. the noise caused by the coaxial gas jet around the beam, were negligible compared with the emissions produced by the surface changes in the metals studied. For plain carbon steels a good correlation was obtained between the extent of the heat-affected zone, in which the transformation to the hard phase martensite had taken place, and the ring-down count (Fig. 17). Consistent with this interpretation of the AE was that stainless steel, which did not undergo the martensitic transformation, gave few counts. The authors suggested that the AE signals could be used as part of a feedback loop to control the extent of the transformation hardening.

In powder fusion the count rates and the RMS voltages resulting from the powdered alloy being fed on to the substrate were high and swamped any emissions associated with the melting or solidification of the hardfacing deposit. It follows that AE cannot be used to assess the quality of the hardfacing deposit while the feed is in operation. On the other hand, RMS voltage may be a convenient method for monitoring the feed *per se* and other production variables such as ultrasonic vibration of the substrate. AE was shown to be suitable for monitoring the deposit during cooling after the feed had been stopped. Cracking during cooling of a Stellite 6 deposit on steel was detected and a reasonable linear relationship between the

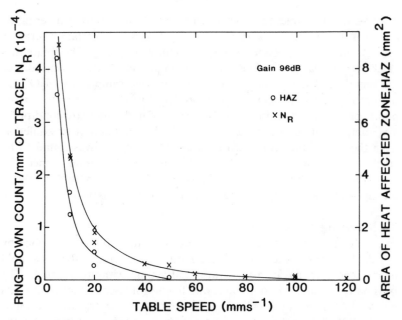

FIG. 17.   Graph showing the similarity between the table speed dependence of the area of the heat affected zone, in which martensite is formed, and the ring-down count for En8 steel.[62]

accumulative event count and the total length of defect (pores plus cracks) was obtained.

A metal may be selected for a particular component because it forms a protective oxide coating under service conditions. For example, a chromium oxide layer of only a few micrometres in thickness ensures good resistance to hydrogen permeation and consequently metals which have a chromium(II) dioxide coating are used in applications such as steam reformers for substitute natural gas synthesis. As described previously for spray coatings, naturally occurring protective oxide layers may also suffer damage because of rapid temperature changes which can impair their protective capabilities. Acoustic emission has been employed to quantify thermal cycling damage, which was not detectable by post-test metallographic examination, in the oxides on a number of metals.[63] In addition, the effect of the damage on hydrogen permeation was measured by the parameter $U = d\phi/\phi_{OM}$, where $d\phi$ is the change in the rate of permeation as a result of thermal cycling and ϕ_{OM} is the original permeation flow through the undamaged oxide. The oxide coating on a metal was considered to

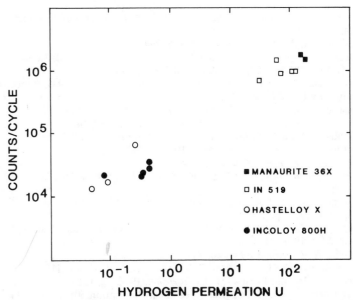

FIG. 18.   Graph showing the correlation between the AE during thermal cycling
and hydrogen permeation measurements.[63]

exhibit good cycling resistance if $U < 1$ and poor resistance if $U > 10$; the
number of counts from a satisfactory coating was found to be an order of
magnitude less than that from a coating susceptible to damage (Fig. 18).

Other coatings which have been studied by means of AE include
platinum aluminide diffusion coatings on a high-temperature nickel alloy[64]
and coatings consisting of glass or ceramic particles in an organic carrier
for oxidation protection.[65]

REFERENCES

1. BEATIE, A. G., *J. Acoustic Emission*, 1983, **2**, 67, 69.
2. HAMSTAD, M. A., *J. Acoustic Emission*, 1983, **2**, 57.
3. BRINDLEY, B. J., HOLT, J. and PALMER, I. G., *NDT Internat.*, 1973, **6**, 299.
4. STONE, D. E. W. and DINGWALL, P. F., *NDT Internat.*, 1977, **10**, 51.
5. ROOUM, J. A., Ph.D. Thesis, University of London, 1980.
6. HOLT, J. and EVANS, T. W. M., Central Electricity Research Laboratories
   Research Report No. RD/L/N40/76, 1976.
7. ONO, K., *Mat. Eval.*, 1976, **34**, 177.
8. HOLT, J. and PALMER, I. G., *Schallemission Symposium, Deutsche Gesellschaft
   für Metallkunde, Munich*, 1974, p. 24.

9. RODGERS, J., *Acoustic Emission Trends, Acoustic Emission Technol. Corp.*, 1982 **2**(3), 1.
10. POLLOCK, A. A., *NDT Internat.*, 1973, **6**, 264.
11. POLLOCK, A. A., in *Acoustic and Vibration Progress*, ed. R. W. B. Stephens and H. G. Leventhall, 1973, Chapman and Hall, London, p. 53.
12. GUTENBERG, B. and RICHTER, C. F., *Seismicity of the Earth*, 1949, Princetown University Press, New Jersey.
13. ISHMIMOTO, M. and IIDA, K., *Bull. Earthquake Res. Inst.*, 1939, **17**, 443.
14. PEAPELL, P. N. and TOPP, K., *Met. & Mat. Tech.*, 1982, **14**, 21.
15. HARRIS, R. W. and WOOD, B. R. A., *Metals Forum*, 1982, **5**, 210.
16. SWINDLEHURST, W., *NDT Internat.*, 1973, **6**, 152.
17. CURTIS, G., *NDT International*, 1974, **7**, 82.
18. WADLEY, H. N. G., SCRUBY, C. B. and SPEAKE, J. H., *Internat. Met. Rev.*, 1980, **25**, 41.
19. WILLIAMS, R. V., *Acoustic Emission*, 1980, Adam Hilger Ltd, Bristol.
20. STRIVENS, T. A., private communication.
21. ZAKHAROV, YU. V., REZNIKOV, YU. A., GORBACHEV, V. I. and VASILÉV, V. N., *Prot. Met.*, 1978, **14**, 376.
22. ARORA, A., *Corrosion*, 1984, **40**, 459.
23. CHAKRAPANI, D. G. and PUGH, E. N., *Met. Trans.*, 1975, **6A**, 1155.
24. RETTIG, T. W. and FELSEN, M. J., *Corrosion*, 1976, **32**, 121.
25. SCHOFIELD, B. H., ASTM Special Technical Publication 505, 1972, p. 11.
26. SCOTT, I. G., *9th Proc. World Conf. on NDT*, Australian Inst. Metals, 1979, Melbourne, Australia.
27. HANSMANN, H. and MOSLÉ, H. G., *Adhäsion*, 1981, **25**, 332.
28. BAHRA, M. S., STRIVENS, T. A. and WILLIAMS-WYNN, D. E. A., *J. Oil Colour Chemists Assoc.*, 1984, **67**, 113.
29. STRIVENS, T. A. and BAHRA, M. S., *Brit. J. NDT*, 1984, **26**, 344.
30. MOSLÉ, H. G. and WELLENKÖTTER, B., *Z. Werkstofftechnik*, 1978, **9**, 265.
31. MOSLÉ, H. G. and WELLENKÖTTER, B., *Conf. Surface Protection by Organic Coatings, European Federation of Corrosion, Budapest, Hungary, 1979.*
32. MOSLÉ. H. G. and WELLENKÖTTER, B., *Symposium on Acoustic Emission, Deutsche Gesellschaft für Metallkunde, Bad Nauheim, FRG, 1979.*
33. BAHRA, M. S., STRIVENS, T. A. and WILLIAMS-WYNN, D. E. A., *J. Oil Colour Chemists Assoc.*, 1984, **67**, 143.
34. HANSMANN, H. and MOSLÉ, H. G., *Proc. 8th Internat. Congress on Metallic Corrosion, Mainz, FRG*, 1981, p. 1063.
35. HANSMANN, H. and MOSLÉ, H. G., *Adhäsion*, 1982, **26**, 18.
36. MOSLÉ, H. G. and WELLENKÖTTER, B., *Metalloberfläche*, 1979, **33**, 513.
37. HANSMANN, H. and MOSLÉ, H. G., *Proc. 9th Internat. Congress on Metallic Corrosion, Toronto, Canada*, 1984, p. 526.
38. STRIVENS, T. A. and RAWLINGS, R. D., *J. Oil Colour Chemists Assoc.*, 1980, **63**, 412.
39. ROOUM, J. and RAWLINGS, R. D., *J. Mat. Sci.*, 1982, **17**, 1745.
40. MOSLÉ, H. G. and WELLENKÖTTER, B., *Farbe u. Lack*, 1981, **87**, 998.
41. KENDIG, M., MANSFELD, F. and ARORA, A., *Proc. 9th Internat. Congress on Metallic Corrosion, Toronto, Canada*, 1984, p. 73.
42. MILLER, D. and RAWLINGS, R. D., to be published.

43. ROOUM, J. and RAWLINGS, R. D., *J. Coatings Technol.*, 1982, **54**, 43.
44. LEIBINGER, G. M. and MOSLÉ, H. G., *Metalloberfläche*, 1985, **39**, 257.
45. ASTM B117-73 (reapproved 1979).
46. MILLER, D. and RAWLINGS, R. D., *Acoustic Emission and Photo-Acoustic Spectroscopy*, 1983, Institute of Acoustics, London.
47. TAKANO, O. and ONO, K., *Proc. 2nd Acoustic Emission Symposium*, 1974, Japan Indust. Plann. Assoc., Tokyo, p. 6/31.
48. HANSMANN, H., *Ind. Eng. Chem. Prod. Res. Dev.*, 1985, **24**, 252.
49. DE IORIO, I., LANGELLA, F. and TETI, R., *J. Acoustic Emission*, 1984, **3**, 158.
50. WELLENKÖTTER, B. and MOSLÉ, H. G., *Metalloberfläche*, 1981, **35**, 24.
51. MOSLÉ, H. G. and WELLENKÖTTER, B., *Metalloberfläche*, 1980, **34**, 424.
52. MOSLÉ, H. G. and WELLENKÖTTER, B., *Metalloberfläche*, 1983, **37**, 94.
53. STEFFENS, H-D., CROSTACK, H-A. and BECZKOWIAK, J., *Coatings for High Temperature Applications* (proc. of a seminar held at the Joint Research Centre of the Commission of the European Communities, The Netherlands) ed. E. Lang, 1983, Applied Science Publishers, London.
54. STEFFENS, H-D and CROSTACK, H-A., *Proc. 9th Internat. Thermal Spraying Conference*, 1980, Nederlands Institute voor Lastechniek, The Hague, The Netherlands, p. 120.
55. BERNDT, C. C. and HERMAN, H., *Thin Solid Films*, 1983, **108**, 427.
56. SHANKAR, N. R., BERNDT, C. C., HERMAN, H. and RANGASWAMY, S., *Amer. Ceram. Soc. Bull.*, 1983, **62**, 614.
57. SAFAI, S., HERMAN, H. and ONO, K., *Proc. 9th Internat. Thermal Spraying Conference*, 1980, Nederlands Institute voor Lastechniek, The Hague, The Netherlands, p. 129.
58. SAFAI, S., HERMAN, H. and ONO, K., *Amer. Ceram. Soc. Bull.*, 1979, **58**, 624.
59. BERNDT, C. C., in *Ultrastructure Processing of Ceramics, Glasses and Composites*, ed. L. L. Hench and D. R. Ulrich, 1984, Wiley Interscience, New York, p. 524.
60. SHANKAR, N. R., BERNDT, C. C. and HERMAN, H., *Proc. 6th Ann. Conf. on Composites and Advanced Ceramic Materials*, published in *Ceram. Eng. Sci. Proc.*, 1982, **3**, 772.
61. ALMOND, D., MOGHISI, M. and REITER, H., *Thin Solid Films*, 1983, **108**, 439.
62. RAWLINGS, R. D. and STEEN, W. M., *Optics & Lasers in Engineering*, 1981, **2**, 173.
63. JONAS, H., STÖVER, D. and HECKER, R., *J. Acoustic Emission*, 1985, **4**, S78.
64. LEHNERT, G., Cost 50 d/20 Report No. 2, reference given in *Coatings for High Temperature Applications*, ed. E. Lang, 1983, Applied Science Publishers, London, p. 307.
65. CLARK, J. N., *Brit. J. NDT*, 1979, **21**, 312.

CHAPTER 4

Organic Coatings in Corrosion Protection

W. FUNKE

Forschungsinstitut für Pigmente und Lacke eV, Stuttgart, Federal Republic of Germany

1. INTRODUCTION

Organic coatings protect, decorate and identify materials and objects. Of these functions, protection is by far the most important and, also, a prerequisite for the other functions.

The annual costs of corrosion in industrialized countries have been estimated to be about 3% of the gross national product.[1] The Railway Authorities of West Germany have to protect an area of about $40 \times 10^6 \, m^2$, which is about five times the area of Lake Constance! The annual costs for protecting this area by organic coatings amount to about DM50 million.[2]

There are no other methods for material protection more versatile in application than organic coatings.

The use of organic coatings to protect metals, especially iron and steel, may be traced back to the dawn of paint and coatings history. The oldest organic coatings for protective purposes are those of bitumen and tar. They protect by preventing the access of aggressive agents, e.g. water and oxygen, to the coating/support interface and thus constitute the first barrier coatings.

Towards the end of the 19th century the combination of linseed oil and red lead appeared and was introduced industrially. This coating system was used as a base layer to protect iron and steel against corrosion. It soon became obvious that the underlying protective mechanism is more complex than just a simple barrier effect. Linseed oil reacts with red lead to form soaps, which are an important element of film-formation.[3-5] As a result of this reaction, corrosion-inhibiting degradation products such as lead salts of dicarboxylic acids are formed.[6-9] D'Ans and his coworkers[10,11] have also suggested that some electrochemical passivation of the metal surface contributes to protection. Accordingly, red lead was classified as an active, anticorrosive pigment and the protective mechanism as a kind of passivation. Other active anticorrosive pigments were subsequently recommended, e.g. lead cyanamide, lead silicochromate, calcium plumbate, lead tungstate, lead vanadate, barium borate, basic zinc potassium chromate and zinc tetraoxichromate. Most of these pigments did not gain much importance or disappeared from the market after a few years. Their protective mechanism could not always be elucidated satisfactorily. The most prominent of these anticorrosive pigments are 'zinc chromates'. Their protective action comprises both the formation of a protective oxidic layer and electrochemical passivation.[12,13] They were widely applied in industrial corrosion protection and replaced red lead in many cases, until, more recently, aspects of health hazards and environmental protection made the use of zinc chromates questionable.

Zinc phosphate is increasingly used as anticorrosive pigment because of its non-hazardous nature. Though not oxidizing, it may protect by forming insoluble complexes with iron ions.[12]

Zinc dust is a completely different type of anticorrosive pigment. When used in very high concentration this pigment operates at least initially by cathodic protection at defects in coating systems.[12] It is a non-toxic pigment and may be used with advantage for a number of specific purposes.

Film formation in the linseed oil/red lead system occurs very slowly. Therefore other binders were introduced, e.g. special types of alkyd resins. Later on, other physically drying binders like chlorinated rubber and vinyl chloride copolymers were used as substitutes for linseed oil with red lead as pigment, but not always successfully.

All anticorrosive pigments work by means of a small fraction of the pigment which is dissolved in water. Correspondingly, binders used in combination with such pigments should not lock up the pigment particles,[13] but allow some diffusion and access of the soluble fraction to the coating/metal interface. Therefore binders for anticorrosive pigments should allow the diffusion of water and swell slightly. Linseed oil is an excellent example of a binder possessing these properties. The conclusion to be drawn is that the barrier mechanism and the passivation mechanism cannot both be made to work optimally in the same coating film.

The following discussion on the state of the art in corrosion protection by organic coatings will centre mainly on iron and steel but will also be applicable to other metal substrates as far as the coating properties are concerned. Though it is not intended to present a complete debate on corrosion mechanisms it is useful to recall some more important features which are especially relevant to the corrosion of coated metals.

2. CORROSION MECHANISMS OF ORGANIC COATING SYSTEMS

Under atmospheric conditions of high humidity, corrosion of exposed steel by natural water or condensed water usually starts at local corrosion elements. There are different reasons for the existence of small cathodic and anodic areas, e.g. the presence of an incomplete layer of metal oxides, grain boundaries, impurities at the metal surface or differential aeration. In practice the original corrosion elements are not known with certainty.[14] Little is known about the size of such an element but it is generally assumed not to be larger than about 0.01 mm^2.[15] The primary corrosion products depend on the pH of the electrochemical reaction. In the neutral region (pH ca 5–9) iron ions, Fe^{2+}, are formed at the anode, and the cathode is depolarized by oxygen which, together with water, forms hydroxyl anions. In the acid region (pH $\leq$ 5) iron again dissolves anodically and hydrogen is formed at the cathode instead of hydroxyl anions. Though acid corrosion is not yet a common event, the growing acidity of rain allows the reduction of hydrogen ions to occur in some cases.

Primary corrosion products of iron and steel are subjected to subsequent reactions including oxidation, reduction, reoxidation and hydrolysis, until the final product, stable rust, has been formed.[16,17]

Depending on the reaction conditions, especially the pH, a whole series of more or less well defined intermediates may be formed, some of which have very interesting colloidal properties.[18-21] It is important to realize that on both unprotected and coated steel, these intermediates strongly influence the course of corrosion because of their effect on the diffusion rates of reagents and reactants, the polarization of the substrate surface and the phenomenological appearance of corrosion defects. In order to explain mechanisms of paint defects, such as blistering or filiform corrosion, account must be taken of the specific influence of secondary and tertiary corrosion products.

If corroding samples dry out at times, as in cyclic corrosion tests with alternately wet and dry exposure periods, colloidal iron oxide membranes will decay in the dry periods and are built up again during the wet periods at different locations of the metal surface. As these membranes retard or inhibit diffusion processes, partly by their electrical charge, the rate of corrosion decreases because of polarization. Destruction of the membranes during dry periods then allows corrosion to proceed when the samples are exposed to water again. This mechanism explains why cyclic corrosion tests. are considered to be more aggressive than a continuous exposure to humidity as in the salt-spray test.

3. CORROSION STIMULANTS

For any electrochemical reaction, an electrolyte is required. Steel corrodes very slowly when exposed to pure water, because ferrous and hydroxyl ions form $Fe(OH)_2$, which has a low solubility in water (0·0067 g/litre at 20°C), precipitates at the local corrosion elements and inhibits diffusion processes necessary for the reaction. Even in the presence of oxygen, corrosion of steel takes place only at a moderate rate in the absence of an electrolyte.

Certain anions like SO_4^{2-} and Cl^- stimulate the corrosion reaction in addition to serving as electrolytes because they react with steel to form the primary corrosion products, ferrous chloride and sulphate complexes, which are soluble and may diffuse away from the electrode areas. Subsequently, these soluble corrosion products are oxidized, hydrolysed and precipitated at some distance from their origin as rust. The stimulating anion is free again and may re-enter the cycle many times until it becomes locked up somewhere in the insoluble corrosion products.[22-26]

Corrosion-stimulating anions can also suppress the corrosion-inhibiting action of anticorrosive pigments. For example, Cl^- inhibits the anticorrosive action of red lead and zinc chromate. Red lead will bind SO_4^{2-} to form an almost insoluble lead sulphate (0·0425 g/litre at 25°C) and will also react with SO_2 and so neutralize it.

Unfortunately all anticorrosive pigments which may insolubilize SO_4^{2-} or SO_2, e.g. red lead or barium borate, are toxic or hazardous. Moreover, one may question whether it is desirable to allow chemical reactions to occur when they involve the anticorrosive pigment within a coating film or at a coating/support interface of the coating system. Other available solutions of this problem seem more reasonable, e.g. decreasing the permeability of SO_2, or, even better, prevention of the formation of an aqueous phase at the interface.

4. METAL SURFACE PRETREATMENT

There is almost no general statement about corrosion protection by organic coatings which does not emphasize the importance of metal surface pretreatment. Prior to paint application metal surfaces must be clean and suitable for adhesion.

Metal surfaces, after undergoing various production processes, may be contaminated by a variety of impurities[27,28] (Table 1) and even clean surfaces do not always guarantee good adhesion. The purpose of metal surface treatments is to remove surface impurities, corrosion stimulants and osmotically active substances, and to convert surfaces in order to obtain chemical and morphological structures which favour adhesion.

Two kinds of surface pretreatments can be distinguished.

(1) Surface cleaning.
(2) Surface conversion:
    (a) mechanical;
    (b) chemical.

Fatty, oily and greasy impurities may be removed by organic solvents; however, inorganic salts such as sulphates and chlorides as well as other water-soluble substances are not necessarily removed by organic solvents. Therefore aqueous solutions of emulsifiers or detergents containing corrosion inhibitors are often more efficient.

Acid or alkaline pickling are usually very efficient, but afterwards the metal surface is very reactive and needs a thorough water rinse and immediate protection.

Mechanical surface conversion distinctly improves adhesion, probably by providing a good anchoring base for subsequent organic coatings.

TABLE 1
Common Contaminants of Metal Surfaces

Contamination	Source of contamination
Mill scale	Rolling mill
Rust	Transport and storage
Lubricants	Shaping processes
Fats, oils, greases	Metal shaping and cutting; temporary protection
Inhibitors	Temporary corrosion protection
Ions, salts	Atmospheric pollution; natural water
Dirt	Transport and storage; sebacious secretions, etc.

TABLE 2
Steel Surface Pretreatment by Phosphating

Method	Layer thickness (μm)	
Zn/Mn phosphating	1–30	Decreasing
Fe phosphating	< 1	Protective
Phosphoric acid pickling rinse	≪ 1	Efficiency

The most common chemical conversion is phosphating.[29] Several processes are used, which differ in efficiency (Table 2). Phosphating consists of converting metal surfaces to phosphate layers epitactically grown on the substrate.[30,31] Subsequent to phosphating, a passivating rinse with chromic acid is used to inhibit corrosion temporarily at gaps and pores, because crystal growth does not provide a complete coverage of the metal surface.

Phosphate layers are excellent anchoring bases for organic coatings and Neuhaus and his coworkers[30,31] have claimed that they also provide some passivity effect, at least in the case of zinc phosphate.

5.   CORROSION PROTECTIVE MECHANISMS OF ORGANIC COATING SYSTEMS

There are several ways by which organic coatings may protect metals against corrosion:

(a)   Passivation or inhibition by anticorrosive pigments or inhibitors.
(b)   Barrier function by decreasing or preventing the access of corrosive agents to the metal surface.[32-35]

Each protective mechanism depends on specific chemical and physico-chemical properties of organic coatings and their components. Protective coatings are still commonly endowed with more than one protective mechanism in order to increase the efficiency and security of protection. However, before discussing how reasonable this strategy is, the principle and the requirements of each protective mechanism are first briefly recapitulated in order to provide the necessary background for a discussion.

6. PASSIVATION AND INHIBITION BY ANTICORROSIVE PIGMENTS AND INHIBITORS

Corrosion protective coatings are distinguished in most cases from other organic coatings by the incorporation of anticorrosive pigments (Table 3). Various protective mechanisms have been proposed, for example (i) topochemical reactions with and complex formation at the metal surface, (ii) formation of impervious or permeation-retarding protective layers by chemical conversion of dissolved pigment fractions or by reaction with dissolved metal ions, and (iii) electrochemical polarization of local electrodes of corrosion elements following adsorption processes.[12]

In some cases several mechanisms may cooperate; for example, red lead is assumed to protect by soap formation[36] after hydrolytic and oxidative degradation of specific binders, by formation of corrosion-inhibiting degradation products of the binder,[9] by the basicity of the divalent lead component PbO,[37] by formation of protective surface layers at corrosion centres,[11,38] by adsorption processes and by neutralization and by the insolubilization of SO_2 and SO_4^{2-} anions.[39] Lindquist and his coworkers[40,41] have suggested that the reaction $Pb^{2+} + 2e \rightarrow Pb^0$ is the reason for corrosion inhibition. Unfortunately not all suggested protective mechanisms have been convincingly established experimentally. Statements on complex formation are often speculative,[42] and in some instances, e.g. zinc phosphate, the mode of action has still not been clearly demonstrated.[43-45]

The question arises whether and how several corrosion protective mechanisms can be incorporated together in a coating system.

TABLE 3
Important Anticorrosive Pigments

Anticorrosive pigment	Formula
Read lead, minium	$Pb_3O_4 (< 10\% \ PbO)$
Lead cyanamide	$PbCN_2$
Lead silicochromate	$4(PbCrO_4.PbO) + 3(SiO_2.4PbO)$
Dibasic lead phosphite	$2PbO.PbHPO_3.\frac{1}{2}H_2O$
Calcium plumbate	Ca_2PbO_4
Zinc chromate (basic zinc potassium chromate)	$K_2CrO_4.3ZnCrO_4.Zn(OH)_2.2H_2O$
Zinc tetraoxichromate	$ZnCrO_4.4Zn(OH)_2$
Zinc phosphate	$Zn_3(PO_4)_2.2H_2O$

Anticorrosive pigments only protect via their water-soluble fractions or via water-soluble reaction products with specific binders, which means that their anticorrosive properties are only developed in the presence of water. On the other hand, water is consumed in the corrosion of iron and steel. Therefore all measures to prevent the access of water to coating/metal interfaces must interfere with the requirement for water to be available for anticorrosive pigments to operate. Because of these conflicting conditions doubt must be cast on whether within the same coating layer corrosion protection by anticorrosive pigments is compatible with other protective mechanisms and whether two or more mechanisms can operate optimally in the same layer.

A major disadvantage of anticorrosive pigments, except zinc phosphate and zinc dust, is that they present health hazards. Handling, application, storage and disposal have been subjected to strict regulations, which greatly restrict the usage of anticorrosive pigments. Moreover, during the last 15 years, it has become increasingly obvious[46,47] that corrosion-protective coatings containing anticorrosive pigments failed because of the corrosive and corrosion-stimulating action of atmospheric pollutants, like SO_2, and anions such as SO_4^{2-}, NO_3^- or Cl^-—the latter one from deicing salt.

This situation provides an impetus for the search for suitable solutions for effective and environmentally safe corrosion protective coatings.

7. CORROSION PROTECTION BY ZINC-RICH ORGANIC COATINGS

Several mechanisms have been proposed to explain the corrosion-protective properties of zinc-rich coatings. The zinc particles act as sacrificing anodes which must be closely packed in order to make the film conductive.[48-51] This electrochemical mechanism operates at least initially until the zinc particles become covered with zinc oxide.[52] Zinc is also able to neutralize acid agents.

Unfortunately most reaction products of Zn and ZnO with common inorganic acids are readily soluble in water. Oxidation of zinc dust to ZnO and transformation to basic carbonates have been assumed to make the coating denser.[53] Some types of zinc dust have a plate-like structure and thus also contribute to the barrier function of the coating.[54]

An advantage of zinc-rich coatings is their non-hazardous character. However, the fields of application are somewhat limited. As in normal cathodic protection, subsequent coatings should shield the zinc-rich layer against all aggressive agents, especially water. Such agents react with zinc dust below the intact covering layers and transform the metal to ZnO and basic zinc carbonates, or worse in this case, to water-soluble zinc salts. The zinc-rich basecoat should only operate at defects of the coating system which extend down to the primer or even to the metal surface.

Similar considerations hold true for coating systems applied to a zinc layer. Intermediate coats and topcoats should protect these base layers against water and other aggressive agents which may react with the zinc. Zinc should be only sacrificed at defects in the coating in order to protect the metal cathodically. Such coating systems do not give 'twofold' protection against corrosion, but the organic coating layers above the anticorrosive base layer have to ensure that its anticorrosive function become effective only where it is needed.

As a consequence of these considerations topcoats and intermediate coats, besides possessing other properties, should be excellent barriers.

8. THE BARRIER MECHANISM AND POROSITY

An important protective mechanism of organic coatings is to bar corrosive agents from the metal surface. This function is referred to as the barrier mechanism.[32,55,56] Water and oxygen are reactants for the primary corrosive reaction. In addition, water also serves as a medium for the electrolyte. Sulphur dioxide may participate directly in the dissolution of iron, and anions, e.g. sulphate and chloride, are both charge carriers and corrosion stimulants. Finally, for balancing electrical charges, cations such as Na^+, NH_4^+, Ca^{2+} or Mg^{2+} may be required.

Water and oxygen have to be supplied continually in sufficient amounts in order for the corrosion reaction to commence and continue. Anions and cations, if not already present at the metal surface before it has been coated, have to diffuse to local corrosion centres.

To judge the efficiency of coating systems as barriers, the diffusion or permeation rates of corrosive agents, the mechanism of transport and the pathways of diffusion must be known. Corrosive agents may take several pathways to arrive at their reaction site (Fig. 1).

An important feature of coating morphology is porosity. As far as diffusion processes are concerned, small and large pores have to be distinguished.

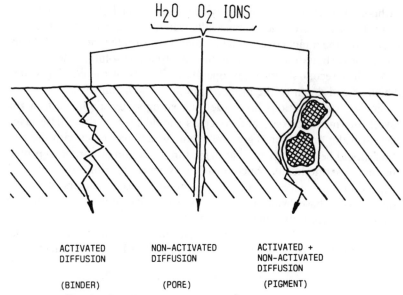

FIG. 1.   Possible types of diffusion in organic coatings.

8.1. Small Pore Systems

If the sizes of pores are within the range 1–5 nm or smaller, they compare with the average 'jump distance' (the distance to be covered by a molecule moving from one minimum of energetic potential to another in the neighbourhood by diffusion) of diffusing molecules or ions. In this case the question arises whether the coating should be considered as a continuous one-phase system, in which the chain segments of the binder participate in intimate molecular, energetic or frictional interactions with the permeating species.

At temperatures above the glass transition temperature of the binder such small pores continuously arise and disappear in the binder matrix due to Brownian motion of the polymer chain segments. The temperature dependence of the permeation rate reflects the temperature dependence of the chain mobility and not that of the viscous flow of the permeating species. This diffusion process is called activated diffusion.

Flexible polymer chains, e.g. those of elastomers, favour activated diffusion. An increase of the crosslink density decreases the rate of activated diffusion, because the segmental mobility of the polymer chains is decreased.

Below the glass transition temperature small pores are frozen-in, and it is questionable whether they are really pores of some length or rather voids within the binder matrix.

As the diffusional pathways are within the range of molecular dimensions it is justifiable to define the morphological structure of such coatings in terms of a homogeneous or one-phase system.

8.2. Large Pore Systems

Organic coatings with pores much larger than the mean jump distance of diffusing molecules or ions are more adequately defined as two-phase systems. These pores are static elements of the morphological structure and their geometry is largely independent of the temperature. The temperature dependence of the diffusion rate corresponds approximately to that of the viscous flow of diffusing species. In contrast to the kinetic porosity of small-pore systems we have, in this case, a static structural porosity.

Diffusion usually takes place through the binder of the coating section at the beginning of the exposure. If subsequently water has interacted at the interfaces between the binder and the pigment and between the coating and the metal surface, then diffusion can occur in these regions also.

Transport of corrosive agents through pigment-free organic coatings depends on various parameters such as:

(1) Structure of the binder—flexibility of the polymer chains, crosslink density, film morphology, ionic charges.
(2) Temperature—experimental temperature, T, below or above the glass transition temperature, T_g, of the binder.
(3) Size and concentration of diffusing species.

9. WATER DIFFUSION

Several pathways are available and several transport mechanisms are possible for the diffusion of water through organic coatings. Organic coatings may contain pores or capillaries, which may well bear ionic charges owing to dissociated acid or basic groups of the binder molecules. These charges attract counterions to form ionic double layers. If an electrical potential is applied across the coating section, water may be transported along capillaries by electro-osmosis. For electro-osmosis to occur, pore diameters must be much larger than the thickness of the ionic double layer.

The electro-osmotic flow of water becomes significant only if the diffuse layer of counterions is extensive and the zeta potential, which characterizes the double layer, is high. This situation occurs notably when distilled water is used as flowing liquid. At high concentrations of an electrolyte, the thickness of the ionic double layer decreases and the zeta potential disappears.

Electro-osmosis has been proposed as a mechanism for the transport of water which causes the blistering in the case of pigmented coatings based on linseed-oil bodied binders.[57] A possible driving force is the electrical potential between areas of coating defects or pores acting as anodes and the adjacent coated areas acting as cathodes. However, whether this mechanism contributes significantly to water diffusion in modern binders, which are much less permeable and water-swellable than linseed oil and its modifications, is to be doubted.

Diffusion of water through organic coatings is an important parameter of corrosion protection.[57-60] Water diffuses faster through organic coatings than oxygen and most other corrosive agents.[13,61-62] Indeed, calculations based on the practical corrosion rates of bare steel and the amount of water transported to the coating/metal interface by diffusion show that more water is available than is needed by the corrosion reaction.[13,62]

On the other hand in spite of their permeability to water, organic coatings may protect structures quite efficiently against corrosion over long exposure times. Even very corrosive environments, such as sea-water and some soils, can be resisted by coating systems used for ships, marine constructions or underground pipelines. This practical experience has often been explained theoretically by the presence of anticorrosive pigments in the base layer. Indeed, most conventional corrosion protective coating systems are based on this strategy of having a second line of defence. Some incongruities of this strategy will be dealt with later on.

Finally, we must emphasize that literature reports on the relationship between water permeation and corrosion protection are controversial.[13,63] However, before this relationship can be resolved, the process of water diffusion must be considered in more detail.

Diffusion of water in organic coatings is necessarily accompanied by water absorption, which is usually equated with 'swelling'. Swelling in the thermodynamic sense means solvation of the chain segments of binder molecules by solvents, in this case by water. However, experimental studies show that most of the water taken up by coatings does not swell the binder but accumulates heterogeneously at the binder/pigment[64] and coating/metal interfaces,[65,66] or even at interfaces within the binder itself.

TABLE 4
Equilibrium Water Absorption of Organic Coating Films on
Immersion in De-ionized Water at 23°C

Type of binder	Water absorption[a] (% by mass (based on binder)
Alkyd–Melamine	0·5
Unsat. Polyester	2·1
Epoxy Amide	2·2
Polyurethane	2·7
Cellulose nitrate–alkyd	0·5–1·8
Phenolic, plast.	0·2
Alkyd, linseed oil	2·0
Alkyd, soya oil	1·2
Chlorinated rubber	0·7
Epoxy–coal tar	0·8
Acrylic, waterborne, stoved	0·4

[a] Values have been determined gravimetrically in the laboratories of the Forschungsinstitut für Pigmente u. Lacke eV, Stuttgart; for experimental technique see ref. 65.

Water is a rather poor swelling agent for most organic binders and the amount of water taken up by swelling is low (Table 4).

Accumulation of water at the pigment/binder interfaces may influence water diffusion by favouring the easier non-activated diffusion.[55,67] At higher concentrations, water molecules may displace the binder from the pigment surface and accumulate there, thus creating space for the non-activated diffusion, which is more rapid than the activated diffusion within the binder matrix. The non-activated diffusion shortens the diffusional pathway through the film.

The degree of pigmentation and the concentration gradient of water across the coating film significantly influences water permeability.[68] Normally increasing the pigment concentration is expected to lengthen the diffusional pathway and decrease the permeation rate because pigment particles are impermeable to water (Fig. 2c). However, if the faster non-activated diffusion via water-filled pigment/binder interspaces is possible throughout the film, the water permeability increases with the pigment volume concentration (PVC) (Fig. 2b).

This relationship is particulary relevant if the downstream concentration is close to 100% relative humidity (r.h.), which means approximately under conditions of autodiffusion.

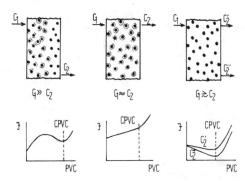

FIG. 2. Influence of activated and interfacial diffusion on permeation rate of water.

In practical permeability testing, a concentration gradient across the coating of 100% r.h. is often chosen. In this case the rate of water permeation is determined for a small layer of the film on the downstream side, within which only activated diffusion is possible. Consequently, in this layer, water is forced to pass through the binder matrix (Fig. 2a). As the PVC is increased the diffusional pathway becomes more circuitous and the permeation rate decreases until it rises drastically at the critical pigment volume concentration (CPVC).

Unfortunately, non-activated diffusion of water is common with pigmented organic coatings at high humidity or on exposure to liquid water. Obviously the pigment/binder interface is rather sensitive to water. Correspondingly, the permeation rate of water through a coating film is also higher at high humidity.[68] Adsorptive bonding between the pigment surface and the binder matrix usually has a polar nature and binder molecules are easily displaced by water molecules. This interfacial interaction of water is not surprising in the case of inorganic pigments. However, even relatively non-polar organic pigments surprisingly tend to accumulate water at their interface to the binder matrix,[66] probably attributable to the surface treatment of such pigments for improving their dispersibility. Displacement of polar bonds by water will be discussed again in the context of wet adhesion.

As the binders used for organic coatings do not swell much, a driving force must operate, which expands the binder matrix in order to provide space for water accumulating at the pigment/binder interface. Besides the well-known clustering tendency of molecular water in relatively unpolar media, osmotic forces may also be involved. Osmotically active substances at the pigment surface or at the pigment/binder interface may be water-soluble fractions or impurities of the pigment or water-soluble substances from the paint adsorbed at the pigment surface before and during film formation.

As may be expected, water accumulation at pigment/binder interfaces is reduced or even prevented when the binder matrix is highly crosslinked and the glass transition temperature is significantly below the practical range of temperature.

The question has frequently been raised whether permeability data obtained experimentally with unsupported coating films are also representative for organic coating on metal supports.[69,70] Of course, if the binder is rigidly bonded to the metal surface by adhesional forces the support may be expected to influence permeability. However, in many cases, as will be shown in the discussion of wet adhesion, water accumulates at the interface after relatively short exposure times and renders the chain segments mobile in the vicinity of the interface. It is therefore reasonable to assume that in presence of water at the interface, molecular relaxations of binder segments are similar to those of the bulk film.

Film-formation of most solvent-borne and oxidatively drying organic coatings does not occur uniformly and inhomogeneities in composition and structure may persist long after paint application.[70,71] Solvents may be retained tenaciously, especially in deeper layers and at the coating/metal interface, even on curing at elevated temperatures. Binders with oxidative film-formation, such as drying oils or alkyd resins, differ significantly in composition and structure through the coating section for a long time after application. When free films of such coatings are prepared, film-formation processes at the underside will be greatly accelerated and will be essentially completed. In this case permeability data of unsupported films are expected to differ from those of supported coatings.

10. DIFFUSION OF IONS

Diffusion of ions in organic coatings is another important aspect of corrosion protection.

Chloride and sulphate anions stimulate corrosion of steel by the immediate formation of soluble complexes of the corrosion products.[72] This reaction prevents local corrosion elements from being polarized by *in situ* formation of insoluble corrosion products. Moreover these ions form an electrolytic solution which is required for corrosion elements to operate.

Similarly to diffusion of water, ion diffusion through organic coatings may take place by an activated diffusion process through the binder matrix or by a non-activated ion migration through virtual pores[73] or internal interspaces filled with water.

Ion diffusion is influenced by the coating's electrical charge which is responsible for the permselectivity.[13,74–78] Binders containing carboxyl or other acid groups, e.g. alkyd resins, are negatively charged in the presence of water and, therefore, are impermeable to anions at low concentrations of the external electrolyte. Correspondingly, positively charged binders, such as epoxy resins cured with an excess of polyaminoamides, are impermeable to cations. Differences found in diffusion rates between cations and anions may be at least partly ascribed to this permselectivity. Accordingly, a number of workers have suggested the use of binders possessing ion-exchange properties for corrosion protective coatings.[79–87]

Whether permselectivity of ions contributes significantly to corrosion protection by organic coatings is still not certain. Apart from the experimental results, to be discussed later on, ion permeability of good protective coating systems is very low and their electrical resistances are very high ($> 10^8$ ohm/cm^2),[88] so that permselectivity may only be expected at very low concentrations of the external electrolyte. Organic coatings based on vinyl polymers have been shown to lose permselectivity at electrolyte concentrations higher than 0.01 N[77] equivalent to a 0.06% (by mass) solution of NaCl. As this concentration is well below concentrations used in salt-spray tests (3–4% by mass), permselectivity cannot play a significant role in preventing permeation of anions through negatively charged coatings under these test conditions.

Maitland and Mayne[86] have amply demonstrated that the electrical resistance of organic coating films may vary with the concentration of an external electrolyte. In I-type films the resistance increases with increasing concentration of the electrolyte. After some weeks of exposure, however, the resistance drops markedly. This drop of resistance was explained by an ion-exchange process, which proceeds rather slowly. In D-type films the resistance decreases with increasing concentration of the external electrolyte, i.e. parallel with decreasing resistance of the electrolyte.

As the resistance of D-type films depends on the electrolyte concentration they do not exhibit a Donnan equilibrium. The lack of a Donnan potential has been interpreted as indicating that there are no pores with charges on the walls.[89] It seems justified, therefore, to assume that such films have no capillaries, at least not above molecular dimensions. As D- and I-type areas occur on the same coating film, Mayne and Scantlebury[90] suggested that the films were heterogeneous in structure, e.g. varying in crosslink density within rather small distances.

Ion conductivity and other electrochemical properties of organic coatings are usually determined with an electrical potential applied externally across the section of the coating. Conclusions on the protective properties drawn from such experiments are therefore based on the assumption that the conductive pathway connecting cathodic and anodic areas runs across the section of the coating. As will be shown later on, however, some evidence exists that this pathway may take its course along the coating/metal interface. In this case ion diffusion through coatings would be irrelevant for underrusting. The irrelevance of ion diffusion through the section of intact coatings would also be consistent with the very low rates of ion diffusion found by radioactive tracers and other techniques, when no external potential is applied.[78,91-94]

Ions must be present at the coating/metal interface in order to provide the conductive electrolyte needed for electrochemical corrosion. These ions may also contribute to osmotic blistering in the course of underrusting. If the primary corrosion products, the ferrous and hydroxyl ions, cannot meet directly after their formation, another counterion must be available to balance the electrical charge. With sodium chloride as an external electrolyte, the counterions for the hydroxyl anions generated at cathodic areas are sodium cations. Other common cations to be considered are those of potassium, calcium and magnesium. As has been stated earlier, two diffusional pathways are available to provide cathodic areas with cations for balancing the electrical charge of hydroxyl ions:

(1)   through the section of the coating; or
(2)   via the coating/metal interface.

With the exception of highly swellable hydrophilic polymers like linseed oil or some cellulose derivatives—not commonly employed for corrosion-protective coatings—binders generally used in paint formulation are efficient barriers to ion diffusion. As there is no water in the coating/metal interface initially, ion concentrations should be higher there than in the external electrolyte. Ion diffusion through the coating section would therefore mean transport against the osmotic pressure. This transport is not very likely unless there are additional factors to act as a driving force, e.g. an external electrical potential.

Recently, a number of workers have shown that cations may migrate in the coating/metal interface.[95-98] This interfacial migration is only possible if an aqueous phase exists there. As a matter of fact, such aqueous phases are formed on the exposure of organic coatings to liquid water or high humidity. The presence of water at this interface manifests itself by a substantial reduction of adhesion.[99,100]

A condition for the interfacial diffusion mechanism is that cations have to enter the interface at some point. It is unreasonable to assume that ion migration occurs via long interfacial pathways when the alternative route is by diffusion through the comparatively thin coating section. Therefore, defects must be present in the coatings, where cations may enter the coating/metal interface.

Funke[100] has shown that alkaline blistering occurs only in the presence of coating defects. Nevertheless, perfect coatings, exposed to sodium chloride solutions, may well develop blisters. However, the aqueous solution in these blisters is neutral. Obviously other mechanisms are responsible for this type of blister formation.[100]

If the electrical charge of hydroxyl anions is balanced by interfacial diffusion of cations to cathodic areas below a defect coating, an explanation is required on how cathodic blisters develop later by osmosis and what the semipermeable membrane is. One explanation could be that after some time positively charged membranes are formed of polymeric complexes of iron hydroxides as the result of the direct reaction between hydroxyl and ferrous ions. Such membranes could prevent the further diffusion of cations to cathodic areas, which are then diluted by water, giving rise to osmotic blisters.

11.   BARRIER PRINCIPLE AND UNDERCUTTING OF DAMAGED COATINGS

An argument frequently advanced in discussions on corrosion protection by coatings is that problems arise with damaged coatings rather than with intact ones and that in such cases permeability to water and oxygen is irrelevant and undercutting becomes important.

There are two different mechanisms for the delamination of coatings on steel, a physico-chemical and an electrochemical one. The physico-chemical delamination, which is more general, is not limited to steel or other metals as substrates. In this case water molecules that have penetrated to the substrate directly disrupt the adhesive bonds between both phases.

More recent results[100-102] indicate that the conclusion that permeability is unimportant for undercutting in the case of damaged coatings must be revised. The reason for this is that the undercutting of the coating adjacent to the damage occurs by cathodic polarization. This polarization

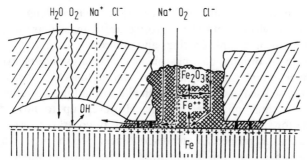

FIG. 3. Pathways of diffusion of oxygen and ions in cathodic blistering of defective organic coatings (negatively charged). Dark hatched area: negatively charged collodial rust.

is due to the fact that considerable amounts of oxygen diffusing to damaged areas are consumed by oxidation of iron(II) ions to iron(III) compounds. Accordingly, the damaged area becomes polarized anodically. The oxygen required to depolarize the adjacent cathodic area then has to be supplied mostly by permeation through the coating to the substrate. Experimental evidence exists that this pathway of oxygen diffusion is important for blistering.[95]

The same is also true for the transport of water to form blisters adjacent to the damaged areas. Delamination does not occur from the anodically polarized, damaged area since anodic polarization does not cause delamination, but it starts from the cathodically polarized blisters formed beside this area. Therefore delamination of the coating is not a continuous process but occurs stepwise (Fig. 3). This mechanism explains why blisters form at all, adjacent to damaged areas before the coating is delaminated. If the coating has a very low permeability, e.g. arising from a high film thickness, blisters are not formed unless other causes, such as interfacial solvent retention, exist. It is therefore understandable why coatings with the smallest possible water- and oxygen-permeability must be used on steel constructions which are cathodically protected. As cathodic delamination, starting at the damaged area, is impeded by concentration polarization due to hydroxyl anions under the adjacent coating sections, cathodic protection requires binders which are resistant to hydrolysis.

12.   WET ADHESION AND WATER DISBONDMENT

The most important property of organic coatings is adhesion to the substrate. This statement is especially true for corrosion protective coatings. Besides mechanical influences, adhesion is challenged severely by a corrosive environment. The corrosive agent most detrimental to the adhesion both of organic coatings and adhesives is water.[65,66,103,104]

Adhesion of organic coatings on metal surfaces during exposure to liquid water or high humidity has been referred to as 'wet adhesion'.[102,12] As 'wet' is also used in connection with film formation, 'wet adhesion' may give rise to some misunderstanding. However, up to now, no better substitute for this term has been proposed. In contrast to wet adhesion, which is a coating property, water disbondment[125] is the process by which a coating is separated from its substrate by the action of water.

Several workers have amply demonstrated that water may displace adhesional bonds at the coating/metal interface.[65,104-106] Industrial metals like steel or aluminium are covered with thin layers of oxides. The thickness of these natural oxide layers is 10–30 nm[107-109] in the case of steel and ca 10 nm in the case of aluminium.[110] Therefore, the structure and composition of these oxide layers have to be considered in the discussion of adhesion of organic coatings. Several theories on the nature of bonding polymers to metal surfaces exist. In case of steel it is generally assumed that polar or acid–base[130] interactions between the binder molecules and the metal oxide surface are responsible for adhesion. Probably hydrogen bonding plays an important role (Fig. 4). More recently in special cases the role of chemical bonding has been discussed.[111-120]

It is uncertain, however, whether the bonds between a metal oxide layer and an organic binder are very resistant to the hydrolytic action of water. Polar bonds are rather sensitive to water molecules, and water disrupts polar bonds between binders and the metal oxide surface, especially if the

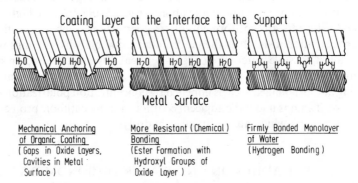

Coating Layer at the Interface to the Support

Metal Surface

| Mechanical Anchoring of Organic Coating (Gaps in Oxide Layers, Cavities in Metal Surface) | More Resistant (Chemical) Bonding (Ester Formation with Hydroxyl Groups of Oxide Layer) | Firmly Bonded Monolayer of Water (Hydrogen Bonding) |

FIG. 4. Simplified schematic of different kinds of interactions responsible for reduced (residual) adhesion on exposure to water.

chain segments bearing the polar groups are mobile and the polar interactions are dynamic. It is therefore very likely that in the absence of mechanical bonds, the failure in the disbonding of wet coatings is essentially an adhesive one,[63,104,106] contrary to the cohesive failures observed with dry films.[121,122]

For this interaction of water at the interface to become effective, the water concentration, to which the coating is exposed, must be above a certain limit.[106] Obviously, clustering of water molecules at the interfaces of organic coatings is only possible at high humidity or on exposure to liquid water. In this case, adhesion is reduced within a comparatively short time to a small fraction of what is observed with the dry coating.[103,104,106,123] Some authors consider that this decrease of adhesion on exposure to water is the first step to underrusting of organic coatings.

Physical data on measuring wet adhesion indicate rather low values as compared with data obtained by adhesion measurements in the dry state.[103,106] Nevertheless, this reduced or residual adhesion is still sufficient to allow blistering before the organic coating will show underrusting and finally is detached from its substrate. The mechanism by which a coating is disbonded by water has been recently discussed.[125] Besides the direct attack of water at adhesive bonds, water accumulation at non-bonded areas or at voids in the coating/metal interface has been proposed.

In characterizing wet adhesion, it is more informative to measure the time required to reduce the strength of the adhesive bond to the residual wet adhesion value following exposure to water than to compare data from physical measurements such as the pull-off test. Such tests are difficult to perform with wet organic coatings and, as with the corresponding adhesion tests of dry coatings, it is often difficult to distinguish clearly between the various possible modes of failure. Funke[100] has shown that blistering of organic coatings on exposure to water is always preceded by a significant reduction of adhesion. Provided the coating has not been exposed to water for too long a period, then adhesion may be recovered after drying the coating system, though the original adhesion strength is usually not regained.[103,106]

The reduction of adhesion on exposure to water obviously does not lead directly to complete delamination, where no recovery of adhesion is possible. The assumption must be made that this residual adhesion is achieved by some specific bonds which are more resistant to water, e.g. chemical bonds, more firmly bonded monolayers of water in the coating/metal interface (Fig. 5) or mechanical locking. Possibly these residual bonds help to maintain the sites of disrupted bonds of the coating

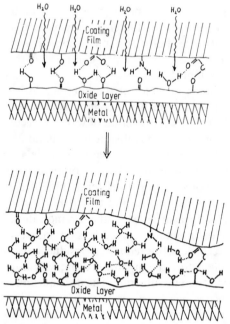

FIG. 5.   Schematic of the coating/metal interface on exposure to water.

at their previous locations on the surface of the substrate, thus assisting recovery of adhesion.

Considering the water-sensitive nature of adhesion, the question is how to make adhesion of organic coatings more resistant to water, other than by preventing its access to the interface. This question is analogous to the problem of making pigment/binder interfaces more water-resistant. In both cases interfacial bonding is susceptible to the interaction of water.

Mechanical bonding to metal surfaces, such as an ink-bottle surface structure, provides an anchoring base for the coating which is largely independent of the water-resistance of the interfacial bond. Sand-blasting is known to improve adhesion by mechanical bonding at the roughened surface.[126] Another example of mechanical anchoring is by phosphate layers. In some cases it is possible to introduce stronger chemical bonding between the coating and the metal surfaces, either directly via reactive groups of the binder or indirectly by adhesion promoters.[127] A general problem is that most chemical bonds are not sufficiently resistant to hydrolysis. A more efficient way to render the adhesion of organic coatings impervious to water is to increase the cooperation of polar adhesive bonds.[32,99] Water molecules can only interfere with polar bonds if the respective chemical bonding groups of the binder are mobile. As these groups are attached to chain segments of the polymeric binder, they can only leave their bonding position if the respective chain segment is mobile. Adhesion forces frequently decrease as the temperatures increase towards the glass transition temperature of the polymer.[128] Above the T_g cohesive failure is observed. As some adhesional bonds may become mobile above the T_g, they are more susceptible to the attack of water than below the T_g, where such bonds are forced to cooperate by their rigid polymer backbone chains.

Relatively rigid, highly crosslinked coatings, for example those for cans or electrical wires, exhibit quite remarkable water resistance despite usually being applied in thin layers. Obviously, in these cases rigid backbone segments force their bonding groups to cooperate in adhesion. Despite polar bonds being susceptible to water, disbonding is only possible if water interferes simultaneously at interfacial bonds over large areas, which is rather improbable.

It would be unreasonable, however, to expect this cooperative principle to apply to coatings of a normal film thickness, because then the completeness of film-formation could suffer and internal stresses as well as inhomogeneities of structure could unfavourably affect other mechanical properties. According to the principle of cooperative bonding, a coating system should have a primer with a thin, highly crosslinked adhesive layer to which is applied a coating layer which may react with chemical groups left at the surface of the adhesive layer during the formation of the film.

13. COMBINATION OF PROTECTIVE MECHANISMS AND CONSTRUCTION OF CORROSION PROTECTIVE COATING SYSTEMS

Organic coating systems for corrosion protection are usually composed of several layers, each of which has a special function. The base layer—the primer—contains the anticorrosive pigment. The second layer—the undercoat—provides a smooth and paintable surface for the top layer, which in addition to possessing barrier properties must contribute to the optical requirements of the coating system, such as colour and gloss. If the basecoat does not cover the surface completely, as is possible with rough surfaces, a second coat of a similar composition is usually applied. Sometimes anticorrosive pigments are also used in this intermediate coat, despite the fact that this layer cannot contribute to passivation.

A correctly formulated basecoat must ensure that a sufficient amount of the soluble pigment fraction diffuses to or is present at areas on the metal surface where neeced. Anticorrosive pigments in the intermediate layer contribute mainly to the barrier function of this layer. In an intact system of this type the top layer and also the intermediate layer must prevent the access of water and corrosion stimulants to the basecoat and especially to the basecoat/substrate interface.

The anticorrosive pigment should only operate at defects of the coating system extending down to the metal support. As anticorrosive pigments have to be soluble in water to some extent, an inclusion of this soluble fraction under less permeable layers may cause an osmotic pressure and give rise to blistering.

For good adhesion metal surfaces must be free of water-soluble impurities before a coating is applied. This requirement must also be true for dissolved fractions of anticorrosive pigments. Therefore, subsequent coating layers have to prevent the access of water to the anticorrosive base layer through the intact area of the coating system and to avoid the dissolution of the anticorrosive pigment there. A combination of passivation and barrier properties in the basecoat is unreasonable. When a protective quality is claimed in this case it is probably to be attributed to the barrier function at the expense of the passivation properties.

Coating systems for external cathodic protection likewise should have optimal barrier properties. They should isolate the metal substrate completely so that cathodic protection is limited to the defect area of the coating system. Otherwise cathodic delamination of the coating system must be expected.

As optimizing both passivation and barrier properties in the same layer is not possible because of different requirements for the binder, the consequences are obvious. The coating system should either be formulated for optimal passivation, taking the risk that it may fail under more corrosive conditions, or the passivation should be dispensed with and the barrier properties made an optimum.

Barrier properties of organic coatings may be greatly improved by choosing plate-like pigments, which by their geometrical shape increase the diffusional pathways through the film[32] and which interact with the binder matrix strongly enough to resist the disbondment by water. By applying this strategy of protection it is possible to develop protective coating systems which may compete with the classical systems and avoid their disadvantages.

An argument frequently advanced is that barrier coatings do not protect

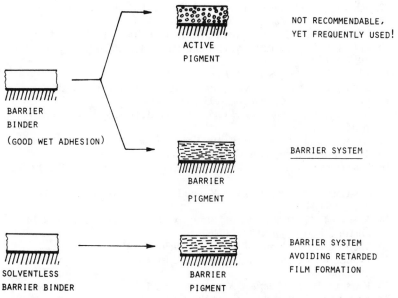

FIG. 6. Variations for improving corrosion protection by organic coatings composed of barrier binders.

at coating defects. However, Funke[102] has shown that for undercutting and underrusting to start from such defects, much depends on water and more especially oxygen permeation through the intact coating adjoining the defect. For optimal protection against corrosion, good barrier properties should be combined with good wet adhesion.[99] This combination will give protection for long periods of continuous exposure to corrosive environments. Figure 6 shows a variety of combinations including those which are not recommended.[129]

REFERENCES

1. Information from the Institute of Science and Technology, University of Manchester (UMIST), Corrosion and Protection Centre, UK.
2. Information from Verband der Lackindustrie e.V., Frankfurt, FRG.
3. RUF, J., Korrosion und Schutz durch Lacke und Pigmente, 1972, Verlag W.A. Colomb, Heenemann GmbH, Stuttgart–Berlin.
4. LAUFENBERG, W., Farbenzeitung, 1929, 35, 546.
5. WEBER, F., Farben, Lacke, Anstrichstoffe, 1949, 3, 329.
6. MAYNE, J. E. O. and RAMSHAW, E. H., J. Appl. Chem., 1960, 10, 419.

7. MAYNE, J. E. O. and RAMSHAW, E. H., *J. Appl. Chem.*, 1963, **13**, 553.
8. MAYNE, J. E. O., *Farbe u. Lack*, 1970, **76**, 243.
9. MAYNE, J. E. O. and VAN ROYEN, D., *J. Appl. Chem.*, 1954, **4**, 384.
10. D'ANS, J., BRECKHEIMER, W. and SCHUSTER, H. J., *Werkstoffe u. Korrosion*, 1957, **8**, 677.
11. D'ANS, J. and SCHUSTER, H. J., *Farbe u. Lack*, 1955, **61**, 453.
12. SVOBODA, M. and MLEZIVA, J., *Prog. Org. Coatings*, 1984, **12**, 251.
13. MAYNE, J. E. O., *Official Dig.*, 1952, **324**, 127.
14. KAESCHE, H., *Die Korrosion der Metalle*, 2nd ed., 1979, Springer Verlag, Berlin, p. 45.
15. KLAS, H. and STEINRATH, H., *Die Korrosion des Eisens und ihre Verhütung*, 1956, Verlag Stahl-Eisen, Düsseldorf, p. 15.
16. STRATMANN, M., BOHNENKAMP, K. and ENGEL, H. J., *Corrosion Sci.*, 1983, **23**, 969.
17. STRATMANN, M., BOHNENKAMP, K. and ENGEL, H. J., *Werkstoffe u. Korrosion*, 1983, **34**, 604.
18. MISEVA, T., HASHIMOTO, K. and SHIMODAIRA, S., *Corrosion Sci.*, 1974, **14**, 131.
19. GRAUER, R., *Chimia*, 1970, **24**, 269.
20. GRAUER, R., *Werkstoffe u. Korrosion*, 1969, **20**, 991.
21. INOUGE, K., TAKANO, T., KANCHO, K. and ISHIKAWA, T., *J. Coll. Interface Sci.*, 1982, **88**, 584.
22. KAESCHE, H., *Werkstoffe u. Korrosion*, 1964, **5**, 379.
23. KNOTKOWA-CERMÁKOVÁ, D. and VLĚKOVÁ, J., *Werkstoffe u. Korrosion*, 1970, **21**, 16.
24. SCHIKORR, G., *Werkstoffe u. Korrosion*, 1964, **15**, 457.
25. DUNKAN, J. R., *Werkstoffe u. Korrosion*, 1974, **25**, 420.
26. BARTON, K. and BERANEK, E., *Werkstoffe u. Korrosion*, 1959, **10**, 377.
27. MAEDA, S., *Prog. Org. Coatings*, 1981, **37**, 28.
28. IEZZI, R. A. and LEIDHEISER, H., Jr, *Corrosion*, 1981, **37**, 28.
29. BENDER, H. S., CHEEVER, G. D. and WOJTKOVIAK, J. J., *Prog. Org. Coatings*, 1980, **8**, 241.
30. NEUHAUS, A. and GEBHARDT, M., *Werkstoffe u. Korrosion*, 1966, **17**, 567.
31. NEUHAUS, A., JUMPERTZ, E. and GEBHARDT, M., *Zeitschr. Electrochemie*, 1962, **66**, 593.
32. FUNKE, W., *Farbe u. Lack*, 1983, **89**, 86.
33. KARGIN, V. A., KARYAKINA, M. J. and BERESTNEVA, Z. YA., *Dokl. Akad. Nauk SSSR*, 1958, **120**, 1065.
34. ORLOV, N. F. and ROZENBLUM, M. YA., *Tr. Ts N II Mor. Flota*, 1973, **175**, 26; *Metal Finish Abstr.*, 1974, **16**, 199.
35. SWEENY, E. E., *Official Dig.*, 1965, **37**, 670.
36. RAGG, M., *Farbenzeitung*, 1929, **34**, 605, 1661.
37. DUNN, E. J., *Treatise on Coatings*, Vol. 3, Pigments, Pt I, ed. R. R. Myers and J. S. Long, M. Dekker, New York, p. 357.
38. PRYOR, M. J., *J. Electrochem. Soc.*, 1954, **101**, 141.
39. LINCKE, G. and ZIELINSKI, TH., *Farbe u. Lack*, 1971, **77**, 443.
40. LINDQUIST, S. A., *J. Oil Colour Chemists Assoc.*, 1984, **67**, 288.
41. LINDQUIST, S. A. and VANNERBERG, N. G., *Werkstoffe u. Korrosion*, 1974, **25**, 740.

42. MEYER, G., *Farben u. Lack*, 1963, **69**, 528.
43. BARRACLOUGH, J. and HARRISON, J. B., *J. Oil Colour Chemists Assoc.*, 1965, **48**, 341.
44. WIENAND, H. and OSTERTAG, W., *Modern Paint & Coatings*, 1984, **74**(11), 38.
45. MEYER, G., *Farbe u. Lack*, 1965, **71**, 113.
46. LANDWEHR, E. and ROSSMÜLLER, H., *Defazet*, 1978, **32**, 234.
47. REICHLE, P., Diplomarbeit, University of Stuttgart, 1986.
48. D'ANS, J. and SCHUSTER, H. J., *Farbe u. Lack*, 1957, **63**, 430.
49. THEILER, F., *Corrosion Sci.*, 1974, **14**, 405.
50. ROSS, T. K. and WOLSTENHOLME, J., *Corrosion Sci.*, 1977, **17**, 341.
51. SCHMID, E. V., *Farbe u. Lack*, 1982, **88**, 435.
52. EBERIUS, E., *Fette, Seifen, Anstrichmittel*, 1958, **60**, 555.
53. PASS, A. and MASON, M. J. F., *J. Oil Colour Chemists Assoc.*, 1961, **44**, 417.
54. BESOLD, R., *Farbe u. Lack*, 1983, **89**, 166.
55. MICHAELS, A. S., *Official Dig.*, 1965, **37**, 638.
56. KUMINS, C. A., *J. Coatings Technol.*, 1980, **52**, 39.
57. KITTELBERGER, W. W. and ELM, A. C., *Ind. Eng. Chem.*, 1947, **39**, 876.
58. HULDEN, M. and HANSEN, C. A., *Prog. Org. Coatings*, 1985, **13**, 171.
59. PERERA, D. Y. and SELIER, P., *Prog. Org. Coatings*, 1973, **1**, 57.
60. BLAHNIK, R., *Prog. Org. Coatings*, 1983, **11**, 353.
61. CORTI, H., FERNANDEZ-PRINI, R. and GÓRNEZ, D., *Prog. Org. Coatings*, 1982, **10**, 5.
62. FUNKE, W. and HAAGEN, H., *Ind. Eng. Chem. Prod. Res. Dev.* 1978, **17**, 50.
63. MCBANE, B. N., *J. Paint Technol.*, 1970, **42**, 730.
64. FUNKE, W., ZORLL, U. and ELSER, W., *Farbe u. Lack*, 1966, **72**, 311.
65. FUNKE, W., *Fette, Seifen, Anstrichmittel*, 1962, **64**, 714.
66. FUNKE, W., *J. Oil Colour Chemists Assoc.*, 1967, **50**, 942.
67. FUNKE, W. and HILT, A., *Angew. Makromol. Chem.*, 1968, **2**, 99.
68. FUNKE, W., *Farbe u. Lack*, 1967, **73**, 707.
69. LAOUT, J. C. and HERRY, F., *Bull. Lab. Professionel*, 1975, **39**, 1.
70. MURRAY, J. D., *J. Oil Colour Chemists Assoc.*, 1973, **56**, 507.
71. KATZ, R. and MUNK, B. F., *J. Oil Colour Chemists Assoc.*, 1969, **52**, 418.
72. SATO, Y., *Prog. Org. Coatings*, 1981, **9**, 85.
73. CHERRY, B. W. and MAYNE, J. E. O., *1st Internat. Congress on Metallic Corrosion, London*, 1961, Butterworths, London.
74. GRUBITSCH, H. and HECKEL, K., *Farbe u. Lack*, 1960, **66**, 22.
75. MAITLAND, C. C. and MAYNE, J. E. O., *Paint Technol.*, 1965, **29**(9), 25.
76. SOLLNER, K., *J. Phys. Chem.*, 1945, **49**, 47, 171, 265.
77. KUMINS, C. A., *Official Dig.*, 1962, **34**, 857.
78. FIALKIEWICZ, A. and SZANDOROWSKI, M., *J. Oil Colour Chemists Assoc.*, 1974, **57**, 259.
79. BARTON, K., ČERMÁKOVÁ, D. and BERANEK, E., *Werkstoffe u. Korrosion*, 1958, **9**, 519.
80. MALIK, W. U. and AGGARWAL, L., *J. Oil Colour Chemists Assoc.*, 1974, **57**, 131.
81. NASINI, A. and OSTACLI, G., *J. Colloid Sci.*, 1956, **11**, 637.
82. SOKOLOVA, E. M., *Lakokras. Mat.*, 1973, **1**, 34.
83. CHERRY, B. W. and MAYNE, J. E. O., *Official Dig.*, 1961, **33**, 469.
84. KUMINS, C. A., *Paint Technol.*, 1964, **28**, 34.

85. CHERRY, B. W. and MAYNE, J. E. O., *Official Dig.*, 1965, **37**, 13.
86. MAITLAND, C. C. and MAYNE, J. E. O., *Official Dig.*, 1962, **34**, 972.
87. ULFVARSON, U. and KHULLAR, M., *J. Oil Colour Chemists Assoc.*, 1971, **54**, 604.
88. BACON, R. C., SMITH, J. T. and RUGG, F. M., *Ind. Eng. Chem.*, 1951, **40**, 473.
89. KINSELLA, E. M. and MAYNE, J. E. O., *Brit. Polym. J.*, 1969, **1**, 173.
90. MAYNE, J. E. O. and SCANTLEBURY, J. D., *Brit. Polym. J.*, 1976, **2**, 240.
91. SVOBODA, M., KUCHYNKA, D. and KNAPEE, B., *Farbe u. Lack*, 1951, **40**, 473.
92. RAO, V. and YASEEN, M., *Pigment & Resin Technol.*, 1978, **7**, 2, 4.
93. MATSUI, E. S., Tech. Note N-1373, February 1975, Civil Eng. Lab. Naval Construction Battalion Center, Port Hueneme, CA 93043.
94. ROTHWELL, G. W., *J. Oil Colour Chemists Assoc.*, 1971, **54**, 992.
95. LEIDHEISER, H., Jr, WANG, W. and IGETOF, L., *Prog. Org. Coatings*, 1983, **11**, 19.
96. FUNKE, W., *Double Liaison*, Oct. 1985, **360**, VIII.
97. KOEHLER, E. L., *Symposium on Corrosion Control by Organic Coatings Lehigh Univ., 1980, Proc. NACE*, 1981, p. 87.
98. COMYN, J., *Polym. Prepr.*, 1983, **24**, 98.
99. FUNKE, W., *J. Oil Colour Chemists Assoc.*, 1985, **68**, 229.
100. FUNKE, W., *Ind. Eng. Chem. Prod. Res. Dev.*, 1985, **24**, 343.
101. FUNKE, W., *Prog. Org. Coatings*, 1981, **9**, 29.
102. FUNKE, W., *J. Coatings Technol.*, 1983, **55**, 31.
103. BULLETT, T. R., *J. Oil Colour Chemists Assoc.*, 1963, **46**, 441.
104. WALKER, P., *Official Dig.*, 1965, **37**, 1561.
105. FUNKE, W. and ZATLOUKAL, H., *Farbe u. Lack*, 1978, **84**, 584.
106. WALKER, P., *Paint Technol.*, 1967, **31**(8), 22; **31**(9), 15.
107. MILEY, H. A. and EVANS, U. R., *J. Chem. Soc.*, 1937, 1295.
108. TÖDT, F., FREIER, R. and SCHWARZ, W., *Zeitschr. Elektrochemie*, 1949, **53**, 132.
109. SCHWARZ, W., *Zeitschr. Elektrochemie*, 1951, **55**, 170.
110. ALTENPOHL, D., *Metall*, 1959, **9**, 164.
111. LEIDHEISER, H., Jr, MUSIC, S. and SIMMONS, G. W., *Nature (London)*, 1982, **297**, 667.
112. PLUEDDEMANN, E. P., *Prog. Org. Coatings*, 1983, **11**, 297.
113. WALKER, P., *J. Coatings Technol.*, 1980, **52**, 49.
114. HOLUBKA, J. W., DE VRIES, J. E. and DICKIE, R. A., *Ind. Eng. Chem. Prod. Res. Dev.*, 1984, **23**, 63.
115. WAKE, W. C., *Polymer*, 1978, **19**, 291.
116. DREYFUSS, P., ECKSTEIN, Y., LIEN, Q. S. and Dollwet, H. H., *J. Polym. Sci., Polym. Lett. Ed.*, 1981, **19**, 427.
117. MITTAL, H. L., *Pure Appl. Chem.*, 1980, **52**, 1295.
118. REINHARD, G., *Plaste u. Kautschuk*, 1984, **31**, 465; 1985, **32**, 32; 1985, **32**, 75.
119. BASIN, V. YE., *Polym. Sci. USSR*, 1979, **20**, 2961.
120. BASIN, V. YE., *Prog. Org. Coatings*, 1984, **12**, 213.
121. VAN OOIJ, W. J., LEIJENAAR, S. R. and VAN DEN BERGH, B., *Fatipec Handbook*, 1982, *Liège*, Belgium.
122. DICKIE, R. A., HAMMOND, J. S. and HOLUBKA, J. W., *Ind. Eng. Chem. Prod. Res. Dev.*, 1981, **20**, 339.
123. BOERS, M. N. M., *Tijd. Oppervlaktetechn. Metalen*, 1977, **21**, 60.
124. FLOYD, F. L., GROSECLOSE, R. G. and FREY, C. M., *J. Oil Colour Chemists Assoc.*, 1983, **66**, 329.

125. LEIDHEISER, H., Jr and FUNKE, W., *J. Oil Colour Chemists Assoc.*, 1987, **70**, 121.
126. ZORLL, U., *Adhäsion*, 1979, **23**, 165.
127. WALKER, P., *J. Coatings Technol.*, 1980, **52**, 49.
128. HUNTSBERGER, J. R., *J. Paint Technol.*, 1967, **39**, 199.
129. FUNKE, W., *Amer. Chem. Soc. Symp. Ser. 322, Polymer Mat. Corros. Control*, 1986, 222.
130. FOWKES, F. M., *Physicochemical Aspects of Polymer Surfaces*, Vol. 2, 1983, ed. K. L. Mittal, Plenum Press, New York, p. 583.

CHAPTER 5

Galvanized Reinforcements in Concrete

M. C. ANDRADE and A. MACIAS

*Institute of Construction and Cement Eduardo Torroja of the CSIC,
Madrid, Spain*

1. THE BEHAVIOUR OF STEEL IN CONCRETE

Concrete is a construction material which has proved to be an excellent protective coating for steel. First of all, it provides a physical barrier between steel and the atmosphere; furthermore it also forms a passive film on the steel because the pH of the aqueous solution filling its pores lies between 12 and 14. At this high pH value the steel is in the passive zone of Pourbaix's diagram.

Numerous reinforced concrete structures are free from corrosion after very long service periods. Nevertheless, many cases have been reported where corrosion of reinforcements has led to destruction of the structure or, at least, the need to repair it (Fig. 1). The two most common factors which produce reinforcement corrosion in reinforced concrete are:

(1) The presence of chloride ions, either added in the mix with the concrete raw materials (water, cement, aggregates or admixtures) or because they penetrate from the outside of the concrete (marine environment).
(2) A decrease of the pH value of the aqueous solution contained within the concrete pores because of the reaction of the cement paste with the atmospheric carbon dioxide (carbonation).

The first factor generally produces localized corrosion and the second a generalized corrosion. The mechanism is of an electrochemical nature. Oxygen and humidity are necessary for the progression of the process.

The consequences of the corrosion of reinforcements are: (a) the decrease

137

FIG. 1.   Example of the effect of the corrosion of reinforcements.

of the bar diameter (weakening its mechanical properties); (b) the spalling and cracking of the concrete cover owing to the expansion associated with the formation of ferrous and ferric oxides; and (c) the decrease of the steel/concrete bond strength.

For many normal environments an adequate design (good quality and thick enough cover, and a correctly detailed reinforcing bar) and the correct preparation of the concrete are enough to insure the durability of the structure. However, for very aggressive environmental conditions, additional protective methods may be necessary.

This chapter deals with one of these complementary protection methods: the galvanization of reinforcements.

2.   PRACTICAL APPLICATIONS OF GALVANIZED REINFORCEMENTS

Galvanizing of reinforcements was introduced to improve the service life of concrete structures in tropical and marine surroundings in the USA.

The first practical use of galvanized reinforcements found in the literature[1] was reported in 1931. The structure was a dock in Bermuda; the hot and wet marine climate of those islands combined with the use of 'coral' aggregates reduced the durability of all reinforced concrete structures. This first experience was successful and led to the use of galvanized reinforcements in such aggressive environments being recommended by the Department of Public Works of Bermuda. Among the numerous examples of buildings and structures constructed there is Bermuda's Airport, which when it was partially demolished after 15 years' service revealed the good performance of the galvanized rebars.[2]

After these encouraging experiences, the use of galvanized rebars was extended in the USA and Canada to bridge deck protection. In the coldest climates of these countries thousands of tons of deicing salts[3,4] (mainly sodium chloride) are utilized to improve the safety[5,6] of roads and motorways. The application of such high amounts of deicing salts leads to the early corrosion of reinforcements in bridges and reinforced concrete pavements. In spite of the use of a thick, dense concrete cover of low permeability the rebars begin to corrode within the first 10 years of life. This deterioration necessitates premature and very expensive[7,8] repair and even rebuilding of these structures at a cost of thousands of millions of dollars.[9]

Numerous bridges have been built in the USA and Canada with galvanized reinforcements.[10,11] Many of them have been inspected some years after erection to check their performance[12,13] and some reports have been published giving the results of these tests.[14,15] Generally, where bare steel was used, there was evidence of corrosion, even in the presence of relatively low amounts of chlorides. However, where galvanized steel was used no evidence was found of significant corrosion or concrete distress.

In spite of these initially encouraging results in the performance of galvanized rebars, the USA Federal Highway Administration (FHA) did not recommend galvanization for general application in bridge decks until more conclusive results were presented[16] because some apparently poor laboratory results were obtained.[17] Three new series of reports[18–20] were supported by the International Lead–Zinc Research Organization (ILZRO) and the American Hot Dip Galvanizers' Association (AHDGA). Again 'Longbird' bridge has been checked after 23 years of life[18] and although the concrete has six times more chlorides than the limit considered to be corrosive, no signs of corrosion were found on its rebars. Eight other bridges erected between 1967 and 1975 in Iowa, Pennsylvania and Vermont were checked in 1981[19] and again only very slight corrosion was found, except in one bridge where a very high water/cement ratio was used. At the National Association of Corrosion Engineers (NACE) *CORROSION-81* meeting, results from the inspection of bridges in

Pennsylvania, Michigan and Florida were presented[20] showing that galvanization always improves performance and lengthens service life.

Other examples of the use of galvanized rebars have been reported[21] including their use in marine environments; for example, some experiments have been done on off-shore platforms. In the ANDOC platform, built in 1976,[22] 2000 tonnes of galvanized rebars were used in the lower tank of the structure.

In the building of the Department of Housing and Urban Development in Washington, DC, finished in 1967, 400 tonnes of galvanized rebars were used to build almost 1600 prefabricated concrete panels.[5] Also, galvanized rebars were used in the US Coast Guard building in Elizabeth City, North Carolina.[5] Among the European examples is London's National Theatre which contains 1000 tonnes of galvanized rebars.[1]

An example of the poor performance of galvanized steel in concrete frequently cited is that given by Mange,[23] who reported the failure of a galvanized sheet and a galvanized mould in contact with concrete which contained a high level of calcium chloride. Also, Arup[24] reports the existence of some failures in Europe but without describing them.

2.1. Laboratory Studies

In spite of favourable practical experiences, the laboratory tests raised controversy over the durability of galvanized steel in concrete.

The best-known earlier papers on galvanized steel in concrete were published by Cornet et al.[25–27] All the results reported the favourable performance of galvanized steel as compared with black steel in chloride-contaminated concrete. However, in 1976, the apparently unfavourable results of Hill et al.[17] moved the FHA to take a more conservative position on the general use of galvanized bridge decks.

Following these first controversial results, numerous laboratory studies have been published and research on the subject remains very active.

Some very good literature reviews on the laboratory studies have been published by Treadaway et al.[28] and Cornet and Bresler.[29] These last two authors commented on the results of Bird,[30–33] Everett et al.,[33] Boyd and Tripler,[34] Duval and Arliguie,[35] Unz,[36] Griffin,[37] Lorman,[38] Clear and Hay,[39] Sopler et al.,[40] Baker et al.,[15] Martin and Rauen,[41] Nishi,[42] Slater et al.,[43] Lewis[44] and Hofsoy and Gukild.[45] Comparison of all these experiments is very difficult because experimental variables and measurement techniques are highly diverse. Although contradictory results can be found, they show that galvanized steel generally improves durability

compared with bare steel in chloride-contaminated concrete; even when galvanized steel fails, bare steel has failed before.
Cornet and Bresler say in conclusion[29]:

(1)   There are no standardized or widely accepted test procedures, making the interpretation of results difficult and often controversial.

(2)   There is general agreement among the reports on corrosion behaviour of galvanized steel that zinc coating furnishes cathodic protection to steel reinforcements: there is less pitting and less intense and extensive corrosion of the steel until substantial quantities of galvanized coating have been consumed.

(3)   As the performance of galvanized steel in concrete depends on the alkali content and on the presence of chromium ion, for the general validity of tests, the variables have to be controlled.

Although this evidence favours galvanized steel, the controversy remains. Moreover, this mistrust of galvanized steel has favoured the use of other alternative complementary protection methods such as epoxy-coated rebars or cathodic protection which, at present, offer a shorter practical service life.

3.   THE MAIN CONTROVERSIAL ISSUES

The main questions leading to the state of confusion are:

(1)   Whether the complementary protection provided by the zinc coating in highly chloride-contaminated concrete is enough.

(2)   Whether zinc is stable in a medium that is as alkaline as the pore solution.

(3)   The possible steel embrittlement caused by hydrogen released during the first few hours of concrete–steel contact.

(4)   Whether the adherence between concrete and reinforcements is impaired and whether it stays above the levels laid down in the standards.

The interest in clarifying this matter has led several research centres to develop very important projects. Among them are the Building Research Establishment (UK), the Laboratoire des Ponts et Chaussées, the Centre Scientifique et Technique du Bâtiment (CSTB), the Centre de Bâtiments et Travaux Publics (CBTP) and the Centre Technique du Zinc (France), the

Max-Planck-Institut für Eisenforschung and the Institut für Baustoffjunde und Stahlbetonbau (FRG), the Norwegian Research Institute (Norway), the Korrosioncentralen and the Danish Technical University (Denmark), the Swedish Cement and Concrete Research Institute (Sweden), the Institutul de Cercetări în Constructii si Economia Construtidor (INCERC) (Romania), the Japan Centre for Construction Materials and the University of Tokyo (Japan) and several Universities and private associations in the USA (American Hot Dip Galvanizers and International Lead Zinc Research Organization), and in Canada, the Ontario Ministry of Transportation.

Also, in Spain the National Metallurgical Research Centre (CENIM) and the Eduardo Torroja Institute of Construction and Cement (IETcc) which belong to the Research Council of Spain (CSIC) have undertaken a joint basic research programme to try to solve the conflicting aspects of the behaviour of galvanized reinforcements. The objectives of this project were related to the various controversial issues and conclusions were drawn from the results obtained which explain the majority of the conflicting points. These are presented in this chapter together with some practical recommendations.

The preliminary studies[46] with cements from different factories and with steel bars from different manufacturers[47] allowed the authors to identify three parameters which decisively influence the corrosion behaviour of galvanized steel in concrete, although these factors are irrelevant for bare steel:

(a)   The cement alkali content which results in different pH values in the aqueous phase of the concrete pores.

(b)   The type of metallographic structure of the galvanized coating which is controlled by the type of base steel (mainly its carbon and silicon content), the bath temperature and time of immersion in the bath.

(c)   The amount of moisture contained in the concrete pores.

As these factors do not influence the behaviour of the bare steel (except in the third case when steel is corroding), they are not usually controlled, which explains the discrepancies found between different studies.

These three factors have been studied extensively by the CENIM and IETcc using synthetic solutions to simulate the pore solution of mortar or concrete specimens. The technique of polarization resistance has been used to measure the instantaneous corrosion rate. Details of the research were published in several papers.[48-55]

4. BASIC ISSUES

4.1. Characteristics of Hot-Dip Galvanized Coating

When the steel to be galvanized is submerged in a bath of liquid zinc (440–460°C), the two metals react in some depth. The coating which remains on the steel, after it has been cooled, has an external layer of pure zinc and several internal layers of iron–zinc alloys, metallurgically linked to the base steel, as shown in Fig. 2. The sequence of layers from the steel substrate to the exterior surface of the coating is:

Gamma layer (Γ)—very thin and usually not discernible. It has a cubic structure with 21–28% Fe.

Delta layer (δ_1)—usually consisting of two layers: one nearer the base steel referred to as 'compact' and one further from it, referred to as a 'barrier'. They have a hexagonal structure with 7–12% Fe.

Zeta layer (ζ)—formed by very asymmetric monoclinic crystals, whose presence diminishes the ductility of the coating. It has 5·8–6·8% Fe.

Eta layer (η)—the external skin consisting of almost pure zinc.

These usual layers have different thicknesses according to the composition of the base steel, the temperature of the bath, the time of immersion and the composition of the zinc bath.[56–58]

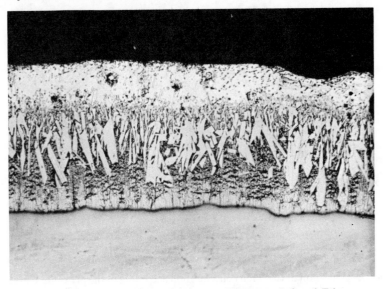

FIG. 2.   Galvanized coating showing the different η, ζ, δ and Γ layers.

The effect of silicon on the reactivity between zinc and steel is known as the 'Sandeling effect'. The influence of silicon is not proportional to its concentration. Thus, when the silicon content of base steel is between 0·5 and 0·15% or more than 0·3% Si the reactivity between the steel and the zinc is higher and the alloyed layers may become considerably thicker.[59,60] The same may happen when the bath has some other alloying elements, such as aluminium.

When the bath temperature is lower than 450°C or the immersion time is longer, the thickness of the external layer of pure zinc increases.

4.2. Nature of the Aqueous Cement Paste Solution

The amount of water mixed with the cement and aggregates to obtain the concrete is always higher than is strictly necessary for cement hydration. This residual water is responsible for the porosity of the concrete and therefore for its permeability. Moreover, all the soluble components of the cement, Ca^{2+}, OH^-, Na^+, K^+ and SO_4^{2-}, go into solution.

The first two ions come from the hydration of the silicates which produce high amounts of calcium hydroxide which saturates the solution. The alkaline ions come from the raw materials and the SO_4^{2-} ions from either the gypsum, added to regulate the cement setting, or from combustible coal or other fuels, which may contain sulphur. The pH value of this aqueous

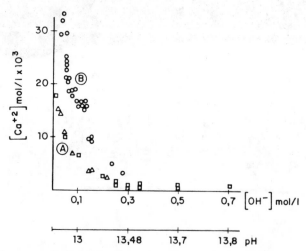

FIG. 3.   Variation of Ca^{2+} ion concentration according to the pH value. Curve A, solutions without SO_4^{2-} ions; curve B, solutions with SO_4^{2-} ions.

solution varies from 12 to 14 as a function of the nature of the cement and the degree of its hydration, which depends on its age.

As is well known, the concentration of Ca^{2+} ions decreases when the pH value increases and it also changes with the ionic strength of the solution. Figure 3 shows the concentration curve of Ca^{2+} against $[OH^-]$, (A) where only $Ca(OH)_2$, NaOH or KOH are present (known as Frattini's curve[61] and (B) where SO_4^{2-} are also considered.[62]

5. INFLUENCE OF THE TYPE OF CEMENT

5.1. Behaviour of Zinc in Very Highly Alkaline Media

Examination of Pourbaix's diagram for zinc (Fig. 4)[63] at the high pH values of the aqueous pore solution, where zinc is in the corrosion zone, demonstrates that, as a first approximation, this metal cannot be considered suitable in such an alkaline medium.

Moreover, looking at Roetheli et al.'s diagram,[64] Fig. 5, zinc is stable below a pH value of about 12·5, but above this value it dissolves at a rate which increases with increasing pH.

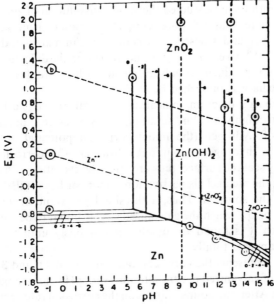

FIG. 4.   Potential pH equilibrium diagram for the zinc–water system at 25°C.[63]

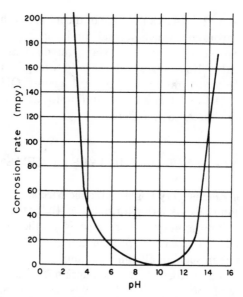

FIG. 5. Corrosion rate of pure zinc as a function of the pH values.[64]

Several authors have studied the anodic dissolution of zinc in solutions which simulate that of pore concrete. Among them are Lieber,[65] Bird,[30] Grauer and Kaesche,[66] Rehm and Lämmke[67] and Duval and Arliguie.[35,68,69] An extensive literature review on the different mechanisms and corrosion products of galvanized steel which are developed in very alkaline media was published by us .[51,54]

As Pourbaix's diagram indicates, when galvanized steel comes into contact with highly alkaline solutions it dissolves very quickly with the evolution of hydrogen (the initial corrosion potential lies below the potential for hydrogen evolution). However, the corrosion products formed may seal the surface, thus arresting the evolution of hydrogen and passivating the metal. This passivation is more or less total, depending on the pH of the solution. Thus, a threshold pH value of $13·3 \pm 0·1$ was established below which the galvanized steel is found to be passivated and above which it is found to dissolve continuously until the galvanized coating disappears totally.

Figure 6 represents the corrosion rates measured 1 h and 33 days after immersion of galvanized bars in solutions with pH values ranging between 12 and 13·6. After one hour's immersion the corrosion rates agree well with the results of Roetheli *et al.*, but after 33 days the solutions which have pH

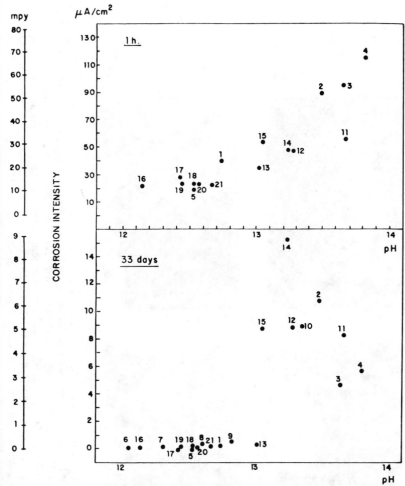

FIG. 6. Variation of the corrosion intensity values of galvanized reinforcements as a function of the pH 1 and 33 days after being immersed in the different solutions.

values below 13·3 show very low corrosion rates (the bars are passivated) and above this limit the corrosion rates remain high enough to lead to the total dissolution of the galvanized coating.

Observing by means of scanning electron microscopy (SEM) the surface of the bars during this process, it was noted that when the pH is around 12·6 the surface is totally covered during the first one or two days by crystallized corrosion products, as shown in Fig. 7(a) and (b). As the pH increases the

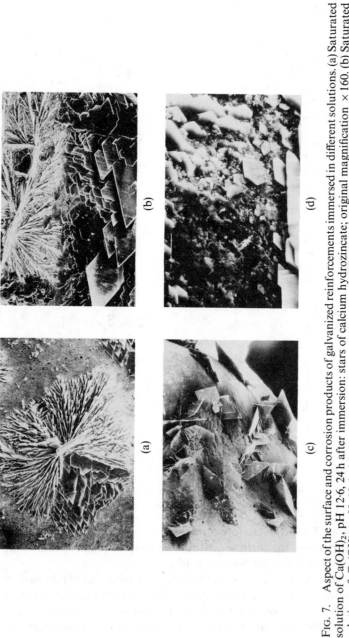

FIG. 7.   Aspect of the surface and corrosion products of galvanized reinforcements immersed in different solutions. (a) Saturated solution of $Ca(OH)_2$, pH 12·6, 24 h after immersion: stars of calcium hydrozincate; original magnification ×160. (b) Saturated solution of $Ca(OH)_2$, pH 12·6, 35 days: overlapping of calcium hydrozincate crystals; original magnification ×152. (c) Saturated solution of $Ca(OH)_2$ + 0·1 M-KOH, pH 12·97, 24 h after immersion: larger calcium hydrozincate crystals; original magnification ×48. (d) Saturated solution of $Ca(OH)_2$ 0·1 M-KOH, pH 12·97, 5 days after immersion: gap of protection between calcium hydrozincate crystals; original magnification ×775.

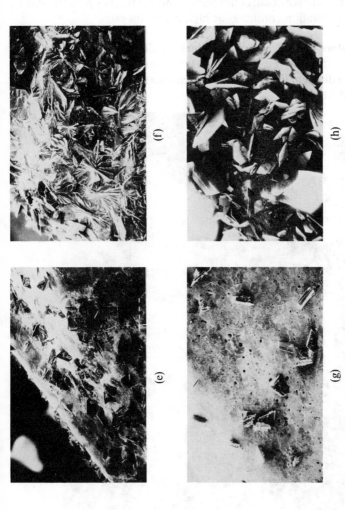

Fig. 7—contd. (e) Saturated solution of Ca(OH)$_2$ + 0·2 M-KOH, pH 13·23, 24 h: some crystals of calcium hydrozincate; original magnification ×51. (f) Saturated solution of Ca(OH)$_2$ + 0·2 M-KOH, pH 13·24, 10 days: calcium hydrozincate crystals not well compacted; original magnification ×12. (g) Saturated solution of Ca(OH)$_2$ + 0·5 M-KOH, pH 13·59, 24 h: isolated crystals of calcium hydrozincate; original magnification ×50. (h) Saturated solution of Ca(OH)$_2$ + 0·5 M-KOH, pH 13·59, 33 days after immersion: calcium hydrozincate crystals, and in the bottom iron oxides and zinc oxide; original magnification ×23.

size of these crystals also increases and they cannot seal the surface so perfectly (Fig. 7c and d), leaving small zones of the metal surface without protection as shown in Fig. 7(e) and (f). Just above the threshold value of 13·3 the corrosion products appear as isolated crystals (Fig. 7g) and thus cannot seal the surface at all; consequently the metal is not passivated (Fig. 7h). Hence dissolution continues at high corrosion rates.

5.2. The Nature of the Corrosion Products of Zinc

As was pointed out by Feitknecht et al.[70,71], zinc oxides and hydroxides exist in many different forms. Thus, the identification of corrosion products of zinc is difficult.

(a)

(b)

FIG. 8.   (a) Aspect of ZnO crystals; original magnification × 2000. (b) Aspect of Zn(OH)$_2$ crystals; original magnification × 156.

Lieber[65] was the first author to identify the nature of the crystals formed on the surface of zinc in contact with a cement paste. He proposed the following mechanism of reaction:

$$Zn + 2H_2O \rightleftharpoons Zn(OH)_2 + H_2 \uparrow \tag{1}$$

$$2Zn(OH)_2 + 2H_2O + Ca(OH)_2 \longrightarrow Ca(Zn(OH)_3)_2 . 2H_2O \tag{2}$$

The product resulting from eqn (2) was identified as calcium hydroxizincate.[35,63,67,73] Also, ZnO (Fig. 8a) and ε-Zn(OH)$_2$ (Fig. 8b) were identified as being formed during the corrosion process in these media.[35,51]

Considering the many results of researchers who have studied the corrosion process in alkaline media, the sequence of reactions which seems best to represent the corrosion process of zinc in these Ca^{2+}-containing media is:

$$\left.\begin{array}{l} Zn + 4OH^- \longrightarrow Zn(OH)_4^{2-} + 2e^- \\ Zn + 2OH^- \longrightarrow ZnO + H_2O + 2e^- \end{array}\right\} \tag{3}$$

$$ZnO + H_2O + 2OH^- \longrightarrow Zn(OH)_4^{2-} \tag{4}$$

$$2Zn(OH)_4^{2-} + Ca^{2+} + 2H_2O \longrightarrow Ca(Zn(OH)_3)_2 . 2H_2O + 2OH^- \tag{5}$$

Figure 9 summarizes the expected behaviour of galvanized steel in solutions with pH values between 11 and 14 concerning morphology of the attack, hydrogen evolution and nature of corrosion products.

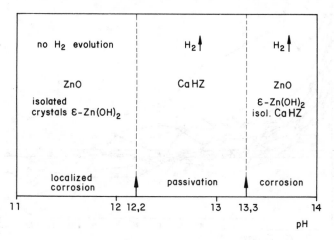

ALKALINE SOLUTIONS CONTAINING Ca^{+2} IONS

FIG. 9. Summary of the behaviour of galvanized rebars immersed in solutions with pH values from 11 to 14. (CaHZ = Ca(Zn(OH)$_3$)$_2$. 2H$_2$O.)

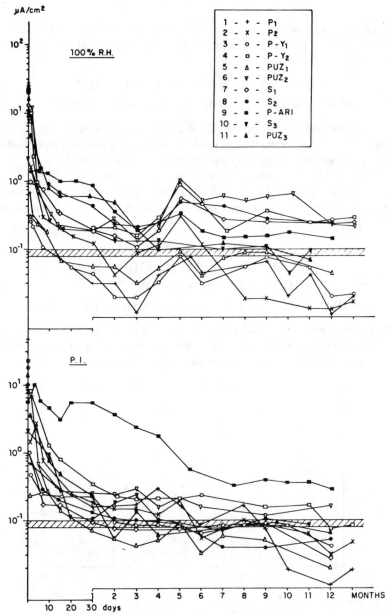

FIG. 10. Corrosion rate of galvanized rebars embedded in mortar fabricated with 11 different cement types and held at 100% relative humidity and partially immersed in water (P.I.).

5.3. Corrosion of Galvanized Bars in Concrete or Mortar

Figure 10 gives a representation of the corrosion rates of galvanized bars embedded in mortar specimens (as shown in Fig. 11) fabricated with 11 different cements and held at 100% R.H. or partially immersed in water (P.I.) The hatched zones in Fig. 10 represent an attack of about 1·05–1·5 μm/year, that is to say a coating of about 60 μm would disappear in about 50 years.

The relation of the alkali content of these cements and the pH of their cement suspensions is presented in Fig. 12. Fig. 10 shows that corrosion rates differ by about one order of magnitude. The cement which suffers least attack (no. 3: P-Y$_1$) is the one with the lowest alkali content. Considering the total corrosion in this first year, a coating of about 60 μm embedded in this cement would last about 200 years, corroding homogeneously. On the other hand, the cement which suffers most attack is no. 9 (P-ARI), which has the highest alkali content. A galvanized coating of about 60 μm would corrode in about 11 years.

Figure 13 shows the relationship between corrosion rates at particular ages and the pH value of the cement suspensions. In spite of the dispersion

FIG. 11.   Mortar specimen used in experiments having two identical rebars as working electrodes.

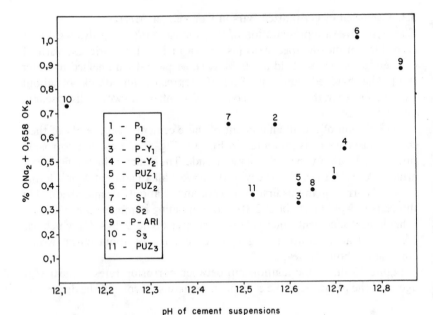

FIG. 12.   Relation between alkali content of the cements (as Na_2O equivalent) of
Fig. 10 and the pH value of cement suspensions (water/cement = 1).

of the values, the pattern of the corrosion rates coincides more or less with
the increasing pH values (or alkali content) of the cements.

This approximate trend between alkali content and corrosion rate is a
very important point in helping to solve the controversy over the stability
of galvanized reinforcements and supports the results found in synthetic
solutions concerning the existence of a threshold value for passivation. This
may explain the fact that as most North American cements have low alkali
contents, owing to the existence of reactive aggregates which provokes
alkali–aggregate reaction in some USA regions, galvanized steel has been
successful there.

As the pH of the solution within the concrete pores is lower than 13·3 in
the first hours following mixing, a layer of passivating corrosion products
(calcium hydroxizincate) will develop and passivate the bar. This protective
layer will be totally passivating as long as the initial pH value is maintained
in the range between ∼ 12·0 and 12·8. Values of between 12·8 and 13·3 may
leave holes in the passivating layer that may continue dissolving when the
pH of the solution later rises because of the disappearance of sulphates in
their reaction with the aluminates.

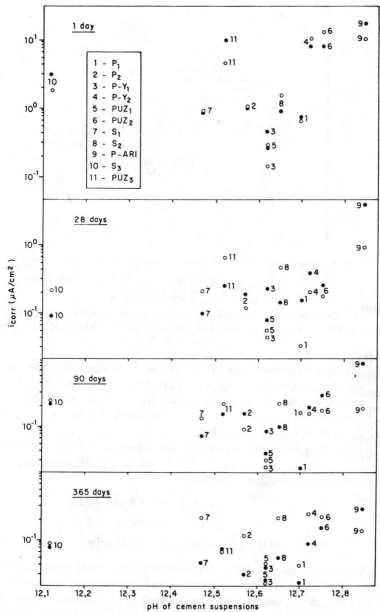

FIG. 13. Relation between the pH value of cement suspensions of Fig. 12 and the corrosion rate of the galvanized rebars on days 1, 28, 90 and 365 of the test.

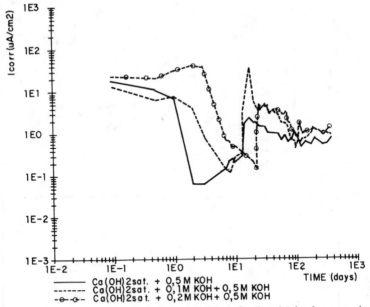

FIG. 14. Corrosion rate as a function of time of galvanized rebars previously passivated in a saturated solution of $Ca(OH)_2$ and then 0.5 M-KOH added (final pH value 13.5).

If the cement solution had a pH higher than 13.3 in these first hours after mixing, the galvanized coating would disappear in the first few days as results in synthetic solutions have shown. However, this possibility never happens if sulphates are used as a setting regulator.[72] While the sulphates are in solution, the pH value always remains just below 13.3; only when sulphates disappear from the solution (as a reaction with the aluminates) does the pH value rise, because redissolution of $Ca(OH)_2$ must occur to maintain the equilibrium of the solution:

$$Na_2SO_4 \longrightarrow 2Na^+ + SO_4^{2-}$$
$$Ca(OH)_2 \longrightarrow OH^- + Ca^{2+}$$
$$Ca^{2+} + SO_4^{2-} + aluminates \longrightarrow ettringite$$

If a continuous passivated layer develops during the setting period a further increase of the pH value does not affect its stability, as tests have demonstrated.[54] These tests are summarized in Fig. 14, which shows the corrosion rates of galvanized bars previously passivated in saturated solutions of $Ca(OH)_2$ and then submitted to an increase of the pH value by

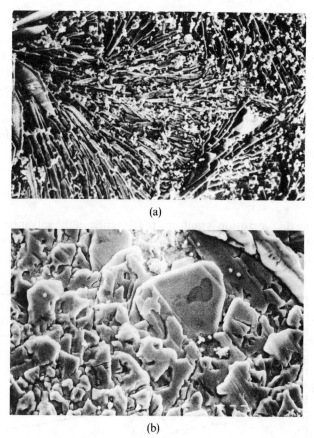

(a)

(b)

FIG. 1. (a) Aspect of the layer of calcium hydroxizincate (CaHZ) 'snowed' by Ca(OH)$_2$ crystals 1 h after the addition of 0·5 M-KOH. Original magnification × 320. (b) Slight dissolution of the borders of the CaHZ crystals 7 days after the addition of 0·5 M-KOH. Original magnification × 306.

the addition of KOH. It may be observed that although the corrosion rate momentarily increases when KOH is added, subsequently the rate reduces to the same level as before. The aspect of the crystals of hydroxizincate 1 h and seven days after the increase of the pH can be observed in Fig. 15(a) and (b). In the former a 'snow' of Ca(OH)$_2$ precipitate can be seen and in Fig. 15(b) a weak dissolution of the crystals of calcium hydroxizincate is apparent.

The whole picture may be summarized as follows: when water is mixed with the cement, the pH value of the solution depends on the alkali content of the cement, but when sulphates are present in the solution, this pH value always remains below the threshold of 13·3. Then when the pH value later rises because the sulphates have disappeared from solution, the surface is already perfectly covered by a protective layer of calcium hydroxizincate, and the durability of the galvanized coating will remain unaltered.

However, this theoretical behaviour is not always reproduced in concrete or mortar because these materials are marked by heterogeneity and other factors, such as the humidity contained in the pores, may produce some variation in the results.

6. INFLUENCE OF THE TYPE OF GALVANIZED COATING

Assuming that corrosion is uniform, the thickness of the galvanized coating necessary for passivating must be about 10 μm. This value is a mean calculated from Fig. 16, representing the shortest (curve A) and longest

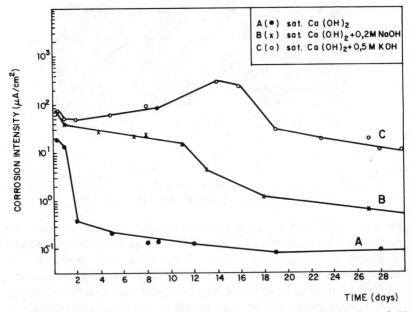

FIG. 16.   Evolution of i_{corr} values of galvanized rebars immersed in solutions of pH values: (A) 12·60; (B) 13·24; and (C) 13·59. Bars in solution A lose 2 μm of their galvanized coating before the CaHZ passivating layer has developed and bars in solution B lose 18 μm. Passivation is not reached by bars in solution C.

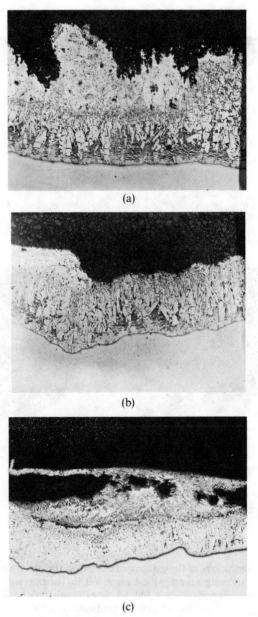

(a)

(b)

(c)

FIG. 17. Different aspects of the typical attack suffered by the pure zinc external layer: original magnifications, (a) ×192; (b) ×200; (c) ×100.

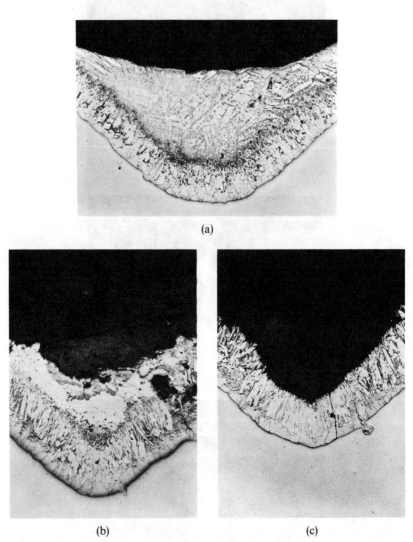

(a)

(b)   (c)

FIG. 18.   Microstructure of the galvanized coating in the bar grooves: (a) the pure zinc fills the 'valley', original magnification ×300; (b) the external layer is partially corroded, original magnification ×300; (c) the external layer has been completely dissolved, original magnification ×300.

(curve B) passivation time for galvanized bars immersed in solutions with pH values of between 12·6 and 13·2. The corrosion intensity measured in case A implies a loss of 2 μm of the galvanized coating before passivation is reached. This loss is 18 μm for curve **B**.

The morphology of the corrosion exhibited by the galvanized coating is shown in Figs 17 and 18. The attack proceeds with dissolution of the external layer of pure zinc (Fig. 17a and b), and development of 'holes' and 'tunnels' (Fig. 17c). The internal layers alloyed to iron do not dissolve. When the bars are corrugated, the morphology of corrosion is similar (Fig. 18a). The amount of zinc in the 'grooves' is totally dissolved before the attack on the other layers starts (Fig. 18b,c).

Therefore, provided the galvanized coating has a sufficient reserve of the pure zinc layer, passivation can be produced without the coating being destroyed. However, when the corrosion rate is very high (solution of pH 13·5), the coating finally disintegrates as shown in Fig. 19.

When the galvanized coating has no external pure zinc layer, as is the case of galvannealed coatings (Fig. 20a), the zinc necessary for developing the protective layer of calcium hydrozincate is removed from the alloying layers which are destroyed (Fig. 20b). The attack is massive and causes

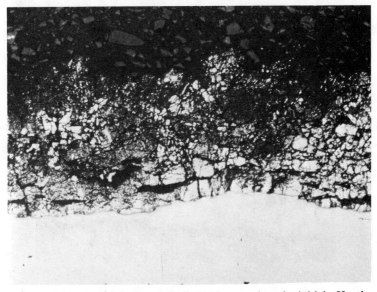

FIG. 19.   Destruction of the internal alloyed layers when the initial pH value is above the threshold; original magnification × 240.

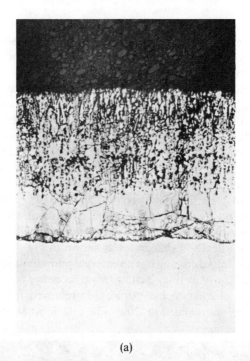

(a)

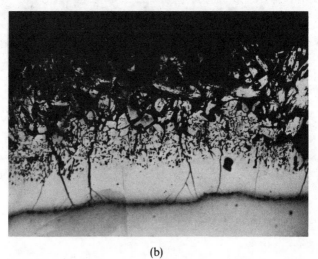

(b)

FIG. 20. (a) Microstructure of a galvannealed coating, original magnification × 300. (b) Type of attack of the galvannealed coating, original magnification, × 160.

disintegration of the coating similar to the case of early dissolution of the pure zinc layer due to high initial corrosion rates (Fig. 19).

Thus, in terms of durability, the most important property of the galvanized coating is the thickness of its external η layer of pure zinc. This layer must have a homogeneous minimum thickness of about 10 μm or, if the thickness is not homogeneous, the coating must have a sufficient 'reserve' of zinc.

7. INFLUENCE OF THE HUMIDITY CONTAINED IN THE PORES

When bare steel is passivated the humidity of the concrete does not affect its passivity. However, when steel is actually corroding, its rate of corrosion is accelerated by increases in the humidity of concrete except, that is, in completely submerged structures, where the access of oxygen to the bars is limited.

With galvanized rebars the changes in concrete humidity may greatly affect the development of passivation during the first days after setting. Afterwards, cyclical changes in the humidity content of the concrete pores may sometimes affect the corrosion rate values dramatically.

The influence of the type of cement and characteristics of the galvanized coating have been well studied but the influence of humidity changes on corrosion rates has not yet been established, and it is not possible to make predictions as to behaviour.

Humidity seems to affect the pH of the pore solution in different ways depending on the type of cement or mortar mix, making the forecast of performance unpredictable. Also, in structures that are not immersed, microcells of differential aeration develop which sometimes increase the corrosion rate by an order of magnitude.

To sum up, dramatic changes in the concrete humidity affect the passivation process of galvanized rebars. Curing of the concrete at a continually high relative humidity is recommended.

8. BEHAVIOUR OF GALVANIZED REBARS IN CHLORIDE CONTAMINATED CONCRETE

Galvanized rebars are really only necessary when the concrete is contaminated by chlorides or has been carbonated, which dramatically reduces the durability of the bare steel. Also, zinc and galvanized steel can support higher amounts of chlorides than bare steel, without pitting.

Chlorides may be present in concrete either because they are added with the raw concrete materials or because they penetrate from the outside.

8.1. Chlorides Added in the Concrete Mix

The presence of Cl^- in the mixing water does not modify the dissolution model described in previous paragraphs. The only difference is that pitting occurs. For the same concentration of Cl^- ion, the susceptibility to pitting decreases as the pH value increases. Thus, a pH range from 12 to 12·6 will allow a higher number of pits than a range from 12·6 to 13·3. Above the threshold of pH of 13·3, the high generalized corrosion which develops masks the pitting phenomenon, which is then undetectable.

The effect of $CaCl_2$ should be different from that of NaCl (or KCl) because $CaCl_2$ decreases the pH of the pore solution and NaCl hardly affects it (NaCl only influences the ionic strength). In fact when $CaCl_2$ is added to mortar the initial corrosion rates are smaller than without admixtures (Fig. 21), but after four to eight months the corrosion rate starts to increase again as pits develop. In tests of longer duration, this increase is not permanent and alternative periods of activation and repassivation occur. Moreover, the number of pits detected on the surface of galvanized

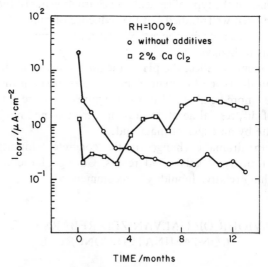

FIG. 21. Corrosion rate of galvanized rebars as a function of time embedded in mortar without admixtures and with 2% calcium chloride held at 100% R.H.

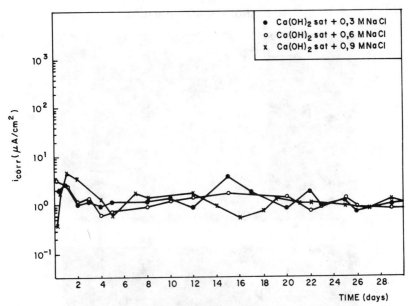

FIG. 22. Corrosion rate of galvanized rebars as a function of time immersed in a saturated solution of $Ca(OH)_2$ with different additions of NaCl.

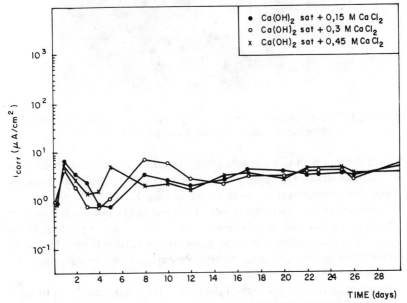

FIG. 23. Corrosion rate of galvanized rebars as a function of time immersed in a saturated solution of $Ca(OH)_2$ with different additions of $CaCl_2$.

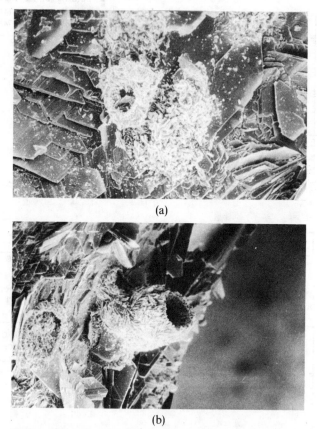

(a)

(b)

FIG. 24.   Different aspects of the pits growing on the outside observed in presence
of chlorides. Original magnifications (a) × 100; (b) × 47.

bars in contact with $CaCl_2$-containing solutions is higher than that found
in NaCl solutions. However, both substances ($CaCl_2$ and NaCl) hardly
change the corrosion rates, as is shown in Figs 22 and 23.

Nevertheless, the most important characteristics of chloride corrosion of
galvanized steel are the curious aspect of the pits and the morphology of the
attack. The pits observed seem to grow on the outside (Figs 24a and b)
rather than into the galvanized coating. Perhaps in concrete this latter type
of growth is physically obstructed.

Concerning the morphology of the attack; the development of pitting is
accompanied by a localized penetration which causes the alloyed layers to
disintegrate (Fig. 25a and b). A preferential dissolution appears to occur

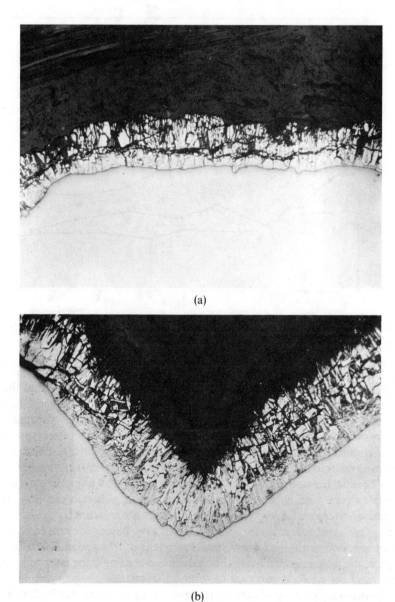

(a)

(b)

FIG. 25.   Typical attack observed in presence of chlorides: the internal alloyed layers have corroded and disintegrated. Original magnifications × 300.

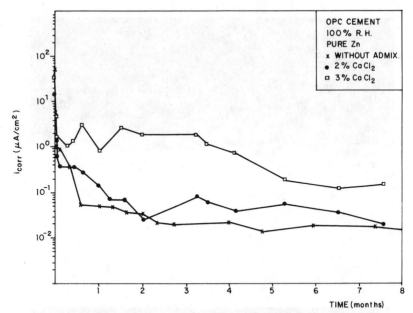

FIG. 26.   Corrosion rate as a function of time, for pure zinc embedded in mortar without admixtures and with 2% and 3% $CaCl_2$, held at 100% R.H.

(some brown oxides are detected) which leads to the total disaggregation of the ζ layer at some particular points. The interpretation may be that the alloyed layers are less resistant to Cl^- attack than the pure zinc layer. This η layer would not cathodically protect the alloyed ones. In the particular points where pitting spreads, the attack may reach the base steel (corrosion rate continuously increases) or may progressively destroy the surrounding coating (alternative periods of relatively high and low corrosion rates).

The most resistant galvanized coatings are those with a thicker external layer of pure zinc. When plain zinc bars are used, although pits may be detected on the surface, very low corrosion rates are measured (Fig. 26). On the other hand, galvannealed coatings are catastrophically destroyed when they are used in chloride-containing concrete (Fig. 27a and b). Therefore, again, the most resistant part of the galvanized coating is the layer of pure zinc and the weakest parts are the alloyed layers.

Concerning the type of cement, it would be expected that the higher the alkali content the lower the susceptibility to chloride attack. However, calcium aluminates (AC_3) and ferrite-aluminates (AFC_4) react with chlorides; hence the real corrosion rate to be detected will depend on the

(a)

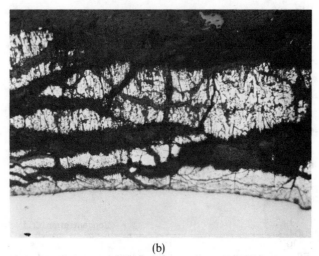

(b)

FIG. 27. Catastrophic and localized attack shown by galvannealed coatings in presence of chlorides. Original magnifications, (a) × 300; (b) × 200.

total concentration of $AC_3 + AFC_4$ as well as the alkali content of a particular cement.

Summing up, galvanized rebars are more resistant to chloride attack than bare steel thus extending the service life of the rebars. However, the extension of service life depends on many factors, including the type of cement, concrete mix proportions, quality and thickness of the pure zinc layer, and humidity content of the pores. Only a careful control of all these factors will ensure that success is achieved, and galvanized rebars become a safe long-term protection method in chloride-containing concrete.

8.2. Penetration of Chlorides from Outside
Following previous statements, it may be deduced that galvanized steel may be a resistant barrier against chloride penetration, but it must have two characteristics when chlorides are present: the protective layer of calcium hydrozincate must be compact and continuous and the remaining coating must be thick enough to resist pitting if it develops (Fig. 28).

Present literature on bridge decks has reported good performance in the majority of cases, even in concrete very highly contaminated with chloride. However, galvanized steel may fail. In these cases, galvanization has always delayed the activated state as compared with bare rebars. The length of this delay will depend on the factors previously mentioned. A cement with

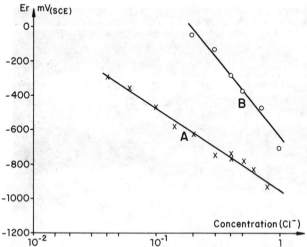

FIG. 28.   Effect of the concentration of chlorides on the pitting potential: (A) pure zinc; (B) pure zinc previously passivated during 15 days in a saturated solution of $Ca(OH)_2$.[68]

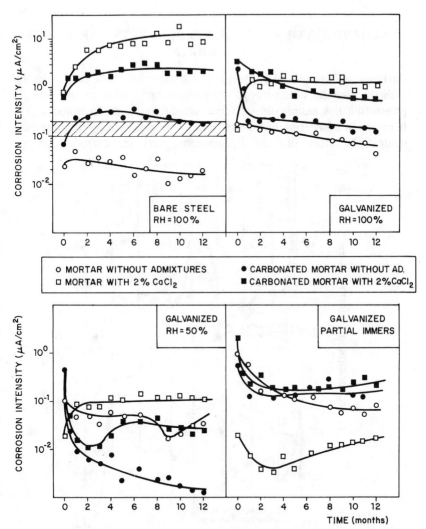

FIG. 29.   Corrosion rate as a function of time for bare and galvanized rebars embedded in carbonated and uncarbonated mortars, without admixtures and with 2% CaCl$_2$.

relatively low alkali content and a galvanized coating thicker than about 100 μm are recommended.

9. BEHAVIOUR OF GALVANIZED REBARS IN CARBONATED CONCRETE

Carbonation is produced by the reaction of the atmospheric carbon dioxide with the alkaline substances of the pore concrete solution and with the phases (silicates, aluminates and ferrite-aluminates) of the hydrated cement. It results in a lowering of the pH value of the aqueous solution reaching neutral pH values. In these circumstances, bare steel corrodes homoge-

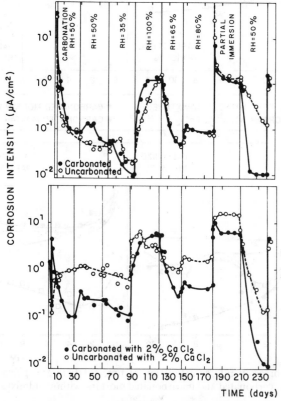

FIG. 30.   Corrosion rate as a function of time for galvanized rebars embedded in carbonated and uncarbonated mortars, without admixtures and with 2% CaCl₂ with cyclic variation of the relative humidity.

neously, producing oxides which lead to the disruption of the concrete cover.

There is no controversy concerning the good performance of galvanized rebars in carbonated concretes. Furthermore, there are recommendations for using galvanized rebars in structures where the cover must be very thin owing to a special design. Galvanized rebars are also recommended in reinforced masonry.

This trust in galvanized steel obviously stems from its superior behaviour in neutral media. However, before the concrete is carbonated, galvanized steel is in contact with the alkaline pore solution and part of the coating is consumed in forming the protective layer of calcium hydrozincate.

Corrosion rates for galvanized steel in mortar specimens with and without chlorides, carbonated in a chamber with about 100% CO_2 and about 60–70% relative humidity, are shown in Figs 29 and 30. Figure 29 shows the behaviour when the external relative humidity was kept constant and Fig. 30 shows the same variables but with cyclic variation of relative. humidity. Even in the presence of chlorides, galvanized steel gives a better performance than bare steel. Galvanization is thus beneficial in carbonated concrete with and without the presence of chlorides.

10.   MISCELLANEOUS PROPERTIES

10.1. Galvanized Steel–Concrete Bond
A bond between galvanized steel and concrete is essential in reinforced concrete and the bond strength depends mainly on the surface pattern of the bar. Thus, plain reinforcements present a lower adherence strength than corrugated ones.

If the coating is chemically inactive, then galvanized bars should have the same level of adherence for the same surface pattern. But galvanized steel reacts in the first hours with the evolution of hydrogen. When the concrete sets, bubbles of hydrogen remain in the bar/concrete boundary, decreasing the effective area of contact between them.

Results on the bond of galvanized reinforcements are controversial. Jokela et al.[73] have reviewed several papers on the matter and their observations are summarized in Table 1. Cornet and Bresler[29] have also published excellent comments on the work of Slater et al.[43] Lewis,[44] Hofsoy and Gukild[45] and Nishi.[42]

Results are contradictory because the type of test (pullout specimens or beam-segmented ones) and the test ages vary from one researcher to

TABLE 1
Galvanized-Concrete Bond Results[73]

Research source	Year	Bar type	Effect of zinc coating on bond
Slater et al.[43]	1920	Plain–rib	Weakens
Schmeer[45]	1920	Plain	Improves
Brodbeck[45]	1954	Plain	Improves
Robinson[45]	1956	Plain (rusty)	Weakens
French research[45]	1959	Plain (rusty)	None
		Plain (pure)	Improves
Bird[45]	1962	Prestressing wire	Weakens
Bresler, Cornet[29]	1964	Plain	None
		Rib	Improves
Gukild, Hofsoy[45]	1965	Rib	Weakens
English research[28]	1969	Plain	Weakens
		Rib	Weakens
Soretz[83]	1971	Rib	Weakens
Maissen[84]	1976	Plain	Weakens
		Rib	None or weakens
Roberts[85]	1978	Plain	None or improves
		Rib	None or improves

another. Also, there is no mention of the alkali content of the cements. Generally, galvanization makes the surface of deformed bars smoother.

Allowing for the variations, the common features drawn from these investigations are:

(1) The development of the bond between steel and concrete is dependent both on age and environment. Hence, the time required for developing a full bond is usually greater for galvanized than for ungalvanized bars.

(2) The bond of galvanized bars may be lower than that of bare steel when both are of the same age but it always reaches the level prescribed by standards.

(3) Chromation of the bars improves the bond of galvanized bars, reaching even higher values than those of bare bars under the same conditions.

A chromating treatment (50–70 ppm of CrO_3 in the mixing water or previous chromatization of the bars[73]) which avoids hydrogen evolution is recommended to avoid the possible decrease of bond in the first months. Figure 31[28] represents polarization curves showing the inhibitive effect of CrO_3 additions. The cements themselves have varying small amounts of

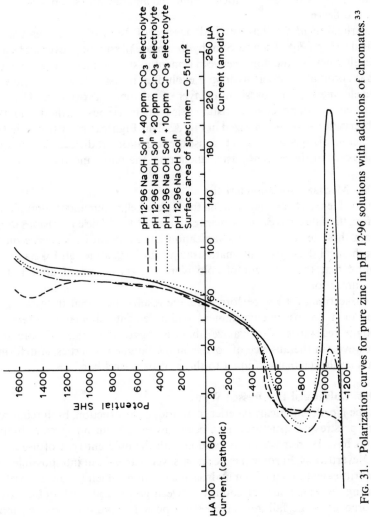

FIG. 31.   Polarization curves for pure zinc in pH 12·96 solutions with additions of chromates.[33]

chromates. This very low chromate content of the cements is another reason for the contradictory results found for the behaviour of galvanized steel in the first hours after mixing: if the chromate content exceeds a critical value then hydrogen evolution is inhibited and the intensity of corrosion is very different.

Arliguie et al.[68,69] have studied the effect of the chromating treatment on the development of the calcium hydrozincate passivating layer and on the development of the bond between cement paste and metal, and concluded that chromates inhibit hydrogen evolution and simultaneously slow the development of the passivating layer of calcium hydrozincate. They also studied the bond characteristics using an ingenious method and their observations are summarized in Fig. 32. This Figure demonstrates that the adherence is age-dependent because the corrosion products diffuse through the pores of the cement paste, 'anchoring' the paste and the metal.

10.2. Mechanical Characteristics
The influence of galvanization on the fatigue behaviour and the mechanical strength of steel have also been investigated.[42,73,74] The conclusions drawn from these tests are that galvanization has no negative influence on the mechanical strength or ultimate strain of the bars, although Jokela et al.[73] found that the 0·2 limit of cold worked bars decreased after hot-dip galvanization.

Concerning fatigue performance, the results show that in tests carried out in a corrosive environment (chloride-contaminated concrete) the fatigue resistance of galvanized bars is higher than that of bare steel. However, in the absence of corrosion the fatigue properties of deformed bars are lower than or equal to those of bare steel bars.

10.3. Galvanized Prestressed Wires
In the North American experience of bridge construction, both reinforced and prestressed structures have been galvanized, but no special observations have been made in connection with the different type of use.

Nevertheless, European experiments were carried out into possible steel embrittlement arising from hydrogen evolution, which has been observed during the first hours of contact between galvanized steel and concrete. Moreover, some failures have been reported because in post-tensioned structures, galvanized grouts were in contact with the steel fibres producing a rapidly propagating brittle fracture in them.

Brachet and Raharinaivo,[75-77] Pini and Hefti,[78] Heiligenstaedt and Bohnenkamp[79] and Riecke[80] have drawn different conclusions concerning

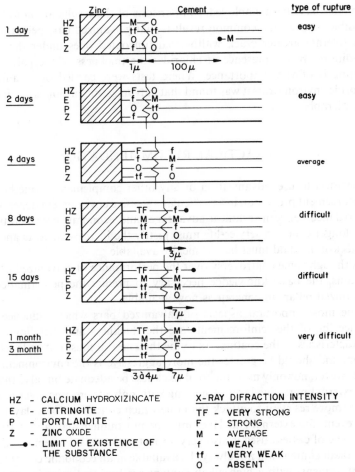

FIG. 32. Characteristics of the galvanized/cement paste interface: type and amount of products and adherence.[69]

the use of galvanization in prestressed structures. Whilst the French have carried out some field trials with galvanized steel fibres, German research workers are very concerned by their use in such types of structures because of the risk of steel embrittlement.

10.4. Behaviour in Cracked Concrete
Only a few experiments have been carried out on cracked concrete-containing galvanized bars.[73,74,81,82] Because very different experimental

variables have been employed, comparisons of the results are not always possible. However, a common result is that galvanized bars perform and can tolerate greater crack widths than uncoated ones under the same conditions. In the presence of chlorides the thickness of the galvanized coating is of vital importance. Where tests were carried out mainly in chloride environments it was found that galvanization prolongs service life considerably.

11. PRACTICAL RECOMMENDATIONS

A decision to use galvanization or any other complementary method of reinforcement protection in a particular agressive environment depends on the balance between technical advantages and economic considerations. The longer expected service life must justify the increase in costs and the protection method must be technically available.

Although some controversy over the performance of galvanized rebars remains, the basic knowledge necessary to predict the performance of galvanized rebars in concrete is now available.

The most important aspects of galvanized bars which influence the service life of the reinforcements are the thickness and metallographic characteristics of the coating. As to the thickness, it must not be less than 60 μm and should be greater the more aggressive is the environment.

However, not only must the total thickness be adequate but also that of the external pure zinc layer. A coating with only a pure zinc layer is very much more resistant to attack than one which contains alloyed layers. In any event, the external zinc layer must be not thinner than about 10 μm. The type of cement used is also very important however; a cement with a low alkali content will not always be available at an economic cost. In any event, cements with a high alkali content are less protective in terms of galvanization but are never aggressive. Moreover, the disparities introduced by the manufacture of concrete make performance less predictable and, therefore, attenuate the importance of the type of cement. However, if different types of cement are available it is recommended to use cements with a low alkali content.

The correct proportioning, fabrication and curing of the concrete are also essential for the good performance of galvanized steel. Since environmental humidity is not easy to control, the most important recommendation is to prevent water draining through the concrete.

REFERENCES

1. PORTER, F. C., *Concrete*, 1976, **8**, 29.
2. ANON., *AHDGA Newsletter*, Jan. 1974, **9**(1), American Hot Dip Galvanizers' Association, Washington, DC 2000 S.
3. SLATER, J. E., *Materials Performance*, 1979, **18**, 34.
4. COOK, H. K. and MCCOY, W. J., in *Chloride corrosion of steel in concrete*, ASTM STP 629, 1976, p. 77.
5. TONINI, E., paper presented at *Corrosion-75*, NACE.
6. PETERSON, P. C., in *Chloride Corrosion of Steel in Concrete*, ASTM STP 629, 1976, pp. 61–8.
7. CAVALIER, P. G. and VASSIE, P. P., *Proc. Inst. Civil Eng*, 1981, **70**(1), 461.
8. CASTLEBERRY, J. R., *Materials Protection*, March 1968, **7**, 21.
9. KILARESKI, W. P., *Materials Performance*, March 1980, **19**, 48.
10. HALL, J. R., *Civil Eng*, Nov. 1964, **34**, 31.
11. ANTONINO, A., *Steel and Corrosion—Some Methods of Protection*, 1967, ASCE, pp. 32–70.
12. COOK, R., paper presented at *Second Annual ILZRO Galvanizing Seminar, June 1976, St Louis, MO.*
13. COOK, A. R. and RADTKE, S. F., paper presented at *Intergalva 76, Madrid, Oct. 1976*; and in *Chloride Corrosion of Steel in Concrete*, ASTM STP 629, 1977, pp. 51–60.
14. STARK, D. and PERENCHIO, W. *The Performance of Galvanized Reinforcement in Concrete Bridge Decks*, Final Report, 1975, Construction Technology Laboratories, Portland Cement Association, Skokie, IL.
15. BAKER, E. A., MONEY, K. L. and SANBORN, G. B., in *Chloride Corrosion of Steel in Concrete*, ASTM STP 629, 1976, p. 30.
16. ANON., *Use of Galvanized Rebars in Bridge Decks*, FHWA Notice N. 5140. 10, July 1976, US Dept of Transportation, Washington, DC.
17. HILL, G. A., SPELLMAN, D. L. and STRATFULL, R. F., *Laboratory Corrosion Tests of Galvanized Steel in Concrete*, Jan. 1976, Report CA-DOT-TL-5351-1-76-02, Calif. Dept of Transportation, Sacramento, CA.
18. STARK, D., *Measurement Techniques and Evaluation of Galvanized Reinforcing Steel in Concrete Structures in Bermuda*, ASTM STP 713, 1980, pp. 132–41.
19. STARK, D., *Evaluation of the Performance of Galvanized Reinforcement in Concrete Bridge Decks*, Project ZE-320, Final Report, Mayo, 1982.
20. SATTERFIELD, D. T. and TONINI, D. E., paper presented at *Corrosion-81*, NACE.
21. ANON., *Berzinkt Staal in Beton Van Buiten.-Gaatse Constructies*, TNO Report 84-2, 1984.
22. GRAFF, W. J. and CHEN, W. F., *ASCE J. Struct. Eng*, June 1981, **107**, 1059.
23. MANGE, C. E., *Materials Performance*, July 1977, **16**, 34.
24. ARUP, H., *Newsletter* (Korrosion Centralen, Denmark), Oct. 1978, No. 1.
25. ISHIKAWA, T., CORNET, I. and BRESLER, B., *Proc. 4th Int. Congress on Metallic Corrosion, September 1969*, pp. 556–9.
26. CORNET, I., ISHIKAWA, T. and BRESLER, B., *Materials Protection*, 1968, **7**(3), 44.
27. CORNET, I. and BRESLER, B., *Materials Protection*, April 1966, **5**, 69.

28. TREADAWAY, K. W. J., BROWN, B. L. and COX, R. N., *Durability of Galvanized Steel in Concrete*, 1980, American Society for Testing and Materials, Washington, DC, pp. 102–131.
29. CORNET, I. and BRESLER, B., in *Corrosion of Reinforcing Steel in Concrete*, ed. D. E. Tonini and J. M. Gaidis, ASTM STP 713, 1980, p. 160.
30. BIRD, C. E., *Corrosion Prevention & Control*, July 1964, **II**, 17.
31. BIRD, C. E. and STRAUSS, J. F., *Materials Protection*, 1967, **6**(7), 48.
32. BIRD, C. E. and CALLAGHAN, B. D., in *Proc. Conf. on Concrete in Aggressive Environments*, Oct. *1977*, South African Corrosion Institute.
33. EVERETT, L. H. and TREADAWAY, K. W. J., *The Use of Galvanized Steel Reinforcement in Building*, Building Research Station Current Paper CP3/70, 1970, Garston, Herts, UK.
34. BOYD, W. K. and TRIPLER, A. B. J. R., *Corrosion of reinforcing steel bars in concrete*. Materials Protection, pp. (40–47), Oct. 1968, NACE, Houston, TX.
35. DUVAL, R. and ARLIGUIE, G., *Mem. Sc. Rev. Met.*, 1974, **71**(11), 719.
36. UNZ, M., *ACI Journal*, March 1978, 91.
37. GRIFFIN, D. F., *The Effectiveness of Zinc Coating on Reinforcing Steel in Concrete Exposed to a Marine Environment*, US Naval Civil Engineering Lab. Tech. Note. N-1032, July 1969.
38. LORMAN, W. R., *Concrete Cover in Thin Wall Reinforced Concrete Floating Piers*, US Naval Civil Engineering Lab. Tech. Note N-1447, July 1976.
39. CLEAR, K. C. and HAY, R. E., *Time to Corrosion of Reinforcing Steel in Concrete Slabs. Vol. 1. Effect of Mix Design and Construction Parameters*. Report no. FHWA-RD73-32, April 1973, US Federal Highway Administration.
40. SOPLER, B., *Corrosion of Reinforcement in Concrete*, Report 73-4, 1973, Cement and Concrete Research Institute, Norwegian Inst. Technology, University of Trondheim.
41. MARTIN, H. and RAUEN, A., *Studies on the Behaviour of Galvanized Reinforcing Steel in Concrete*, Berich no. 68 des Gemeinschaftsausschusses Vezinken eV, Düsseldorf, 1975.
42. NISHI, T., *Investigations on Mechanical Behaviour of Galvanized Steel Reinforcement in Concrete*, ILZRO-Project no. ZE-170, Final Report no. 5, Jan. 1971 to June 1974, Japan Testing Center for Construction Materials.
43. SLATER, W. A., RICHART, F. E. and SCOFIELD, G. G., Technical Paper no. 173, 1920, US National Bureau of Standards, Washington, DC.
44. LEWIS, D. A., in *Proc. 1st Internat. Congress on Metallic Corrosion, London, England, April 1961*, pp. 547–555.
45. HOFSOY, A. and GUKILD, I., *J. Amer. Concrete Inst.*, 1969, **66**, 1974.
46. ANDRADE, C. and GONZALEZ, J. A., *Materiales de Construcción*, 1978, **172**, 71.
47. ANDRADE, C., VAZQUEZ, A. J. and GONZALEZ, J. A., *Revista de Metalurgia*, 1977, **3**, 142.
48. GONZALEZ, J. A., VAZQUEZ, R. and ANDRADE, C., *Revista de Metalurgia*, 1977, **2**, 94.
49. GONZALEZ, J. A., VAZQUEZ, A. J. and ANDRADE, C., *Matériaux & Constructions*, 1982, **15** (88), 271.
50. GONZALEZ, J. A., VAZQUEZ, A. J., JAUREGUI, G. and ANDRADE, C., *Matériaux & Constructions*, 1984, **17**(102), 409.
51. MACIAS, A. and ANDRADE, C., *Brit. Corrosion J.*, 1983, **18**(2), 82.

52. ANDRADE, C., MOLINA, A., HUETE, F. and GONZALEZ, J. A., *Corrosion of Reinforcements in Concrete Construction*, 1983, The Chemical Society, London, pp. 343–55.
53. MOLINA, A., BLANCO, M. T. and ANDRADE, C., *9th Internat. Congress on Metallic Corrosion, Toronto, June 1984*, Vol. 1, p. 412.
54. BLANCO, M. T., ANDRADE, C. and MACIAS, A., *Brit. Corrosion J.*, 1984, **19**(1), 41.
55. ALONSO, C., MACIAS, A., SANTOS, P. and ANDRADE, C., *36th Meeting Interat. Soc. Electrochemistry, Sept. 1985*, p. 6391.
56. MACKOVIAK, J. and SHORT, N. R., *Internat. Metals Rev.*, 1979 (1).
57. RUIZ MARTINEZ, J. L. and VAZQUEZ VAAMONDE, A. J., *Metalurgia y Electricidad (Revista Técnica)*, Feb. 1983 (542).
58. VAZQUEZ, A. J., *Estudio Sobre la Adherencia de las Armaduras Galvanizadas en la Construcción de Hormigón Armado*, Informes de la Construción 258, Mar. 1974, p. 57.
59. SORENSEN, O. B. and MAAHN, E., *Proc. 2nd Internat. ILZRO Galvanizing Seminar St Louis, MO, 9–10 June 1976*, Communication 9.
60. HABRAKEN, L., *Proc. INTERGALVA 79, 12th Internat. Galvanizing Conf., Paris, 20–23 May 1979*, pp. 121–41.
61. FRATINI, N., *Anal. Chimica Applicata*, 1949, **34**, 39, 41.
62. HANSEN, W. I. H. and PRESSLER, F. E. P., *4th Internat. Congress on the Chemistry of Cements, 1960, Washington*, Paper IV-2.
63. POURBAIX, M., *Atlas of Electrochemical Equilibria*, 1966, Pergamon Press, Oxford, pp. 314, 410.
64. ROETHELI, B. E., COX, G. K. and LITTREAL, W. B., *Metals & Alloys*, Mar. 1932, **3**, 73.
65. LIEBER, W., *Zement—Kalk—Gips*, 1967, **3**, 91.
66. GRAUER, R. and KAESCHE, H., *Corrosion Sci.*, 1972, **12**, 617.
67. REHM, G. and LÄMKE, A., *Betonsteinzeitung*, 1970, **6**, 360.
68. ARLIGUIE, G., DUVAL, R. and LONGUET, P., *Ciments, Betons, Plâtres, Chaux*, 1979, **4**, 201.
69. ARLIGUIE, G., GRANDET, J. and DUVAL, R., *7ème Congrés Internat. Chimie des Ciments, Paris*, Vol. III, 1980, VII-22.
70. FEITKNECHT, W., *Helv. Chim. Acta*, 1930, **13**, 314.
71. FEITKNECHT, W., *Helv. Chim. Acta*, 1949, **32**, 2294.
72. MORAGUES, A., MACIAS, A., and ANDRADE, C., *Cement & Concrete Research*, submitted.
73. JOKELA, J., METSO, J. and SARJA, A., *Zinc-Coated Concrete Reinforcements*, Report 13, Dec. 1982, Nordic Concrete Research, Oslo.
74. OKAMURA, H. and HISAMATSU, Y., *Materials Performance*, July 1976, **15**, 43.
75. BRACHET, M. and RAHARINAIVO, A., *Matériaux & Constructions*, 1975, **8**(46), 323.
76. CRETON, B. and RAHARINAIVO, A., *Field Application of Hot-Dip Galvanized Prestressing Steels*, ILZRO Project no. ZE-220, Final Report, Sept. 1983.
77. RAHARINAIVO, A., *Bull. Lab. Ponts & Chaussées*, Nov./Dec. 1980, **110**, 83.
78. PINI, G. C. and HEFTI, B., *Metalloberfläche*, 1978, **32**(8), 329.
79. HEILIGENSTAEDT, P. and BOHNENKAMP, K., *Archiv. Eisenhüttenwessen*, 1976, **2**, 107.

80. RIECKE, E., *Werkstoffe & Korrosion*, 1979, **30**, 619.
81. NÜRNBERGER, V., *Korrosionsverhalten Verzinkter Spannstähle in Gerissenem Beton*, 1984, Deutscher Ausschuss für Stahlbeton, Berlin, **353**, p. 83.
82. SATAKE, J., KAMAKURA, M., SHIRAKAWA, K., MIKAMI, N. and SWAMY, N., *Corrosion of Reinforcements in Concrete Construction*, 1983, The Chemical Society, London, Chapter 21.
83. SORETZ, S., *Betonstahl Entwicke*, 1971, **45**, 8.
84. MAISSEN, A., *Schwiz. Bauztg*, 1976, **94**, 675.
85. ROBERTS, A. W., *Bond Characteristics of Concrete Reinforcing Tendons Coated with Zinc*, Final Report, Univ. Newcastle, New South Wales, Australia, 1978, p. 92.

CHAPTER 6

Titanium Dioxide for Surface Coatings

The late T. ENTWISTLE

Tioxide UK Ltd, Billingham, Cleveland, UK

1. INTRODUCTION

Titanium dioxide is the most widely used pigment in the surface coatings industry, consumption in 1985 being of the order of 1.6×10^6 metric tonnes (Mg) with a total usage in all industries of approx. 2.5×10^6 Mg. Consumption by different industries in various markets is given by Callow.[1]

The predominance of titanium dioxide as a white pigment results from a combination of chemical inertness, high refractive index and an almost complete absence of absorption in the visible portion of the spectrum, which ensures a good whiteness. The importance of these properties will be discussed later in this chapter.

Although titanium is widely distributed and is the ninth most abundant element in the Earth's crust, it was not discovered until 1791 when the Rev. William Gregor, who was the vicar of Menaccan, Cornwall, and also an amateur mineralogist, detected a hitherto unknown metallic oxide present in a locally occurring black mineral sand. He gave the names menaccine to the oxide and menaccanite to the mineral sand after the area of origin. Some years later a German scientist, Martin Klaproth, separated titanium dioxide from a mineral, red schorl, which occurred in Hungary and, unaware of Gregor's discovery, named it titanic earth, after the Titans, the first sons of the earth in Greek mythology. When the two oxides were later discovered to be the same, the name titanium was retained although Gregor was credited with the discovery. The red schorl, used by Klaproth, later became known as rutile and the menaccanite as ilmenite, after the Ilmen mountains in Russia, where large deposits occur.

183

The first recorded commercial use in surface coatings was in Birmingham by Rylands, who, in about 1865, used ground ilmenite as a black pigment; but commercial exploitation of titanium dioxide as a white pigment did not take place until 1913 when it was manufactured by a fusion process. This process was superseded in 1918 by the sulphuric acid or sulphate process—the major process used currently for separation of titanium dioxide from its ore. Manufacturing was carried out by the Titanium Pigment Company at Niagara Falls, USA, and the Titan Company at Frederickstad, Norway. Development work for this process had been started in 1908 by Farup and Jebson in Norway and independently by Rosa and Barton in the USA. In 1920 the two companies agreed on a mutual licensing of patents and exchange of technical information.

These early pigments were poor in colour because removal of ferric iron was incomplete and, compared with present day pigments, had inferior opacity, a consequence of the poor control of crystal size. Satisfactory pigments could only be produced at an economical price by mixing with blanc fixe ($BaSO_4$), a procedure whereby the titanium hydroxide, which is first obtained in solution in the separation process, is precipitated on to the blanc fixe prior to calcination of the hydroxide to titanium dioxide.

Pure titanium dioxide was first produced commercially by Thann et Mulhouse, France, in 1923 using the Blumenfeld process, which was a variant of the sulphate process, but substantial quantities were not available until about 1927.

Titanium dioxide pigments are produced in two crystalline forms, anatase and rutile. A third crystalline form, brookite, also exists but has no commercial application. The earlier pigments were of the anatase form, which is very photoactive, so that paints employing them showed very rapid chalking when exposed to sunlight and moisture. It was known that the rutile form was much less photoactive and, also, its higher refractive index would give higher opacity. However, the technical difficulties in producing a satisfactory rutile pigment were very great. The first commercial rutile pigments were not produced until 1939—in Czechoslovakia—but, because of the Second World War, development in Europe was curtailed until 1945/46. However, work continued in the USA, where rutile pigments were manufactured in the early 1940s.

From the early 1930s to late 1950s, the average annual tonnage increase was of the order of 12% with the number of factories increasing from eight to 28. All used variants of the sulphate process. Many alternative methods of production have been investigated and although many processes are mentioned in the patents, the only one to be commercialised is the 'chloride' process. The first chloride process patent was taken out by DuPont in 1948 but commercial production did not start until 1957/1958. By 1969 approximately 15% of the world's production was manufactured by the chloride process, and by 1985 this had increased to 35%, mainly because in the USA the process accounts for over 90% of production.

2. MANUFACTURE

2.1. Ores

The most important naturally occurring ores are ilmenite and mineral rutile. Ilmenite is a black sand or rock with the theoretical formula $FeO.TiO_2$ ($FeTiO_3$) but the TiO_2 content varies from approximately 40% to 65% depending upon the source. Some of the iron is generally oxidised to the ferric state and in addition there are also significant siliceous impurities. Beach sands contain the higher TiO_2 levels (52–64%) because of the removal of iron by weathering and concentration by wave action. In addition to concentration, weathering results in the conversion of the titanium dioxide to leucoxine, pseudorutile and eventually rutile. Beach sands are commercially extracted in Eastern and Western Australia, South India, Sri Lanka, Malaysia, Florida and South Africa. Significant quantities of the massive form occur in Norway, Finland, Canada, the USA and Sweden.

Mineral rutile is naturally occurring titanium dioxide with a TiO_2 content of the order of 95%. The colour can vary from brown to reddish black with silica and iron being the major impurities. Significant deposits occur in Australia, South Africa and Sierra Leone.

A third naturally occurring ore is a mixed ilmenite/leucoxine/pseudorutile/rutile deposit found in Western Florida and formed by the weathering action on ilmenite.

In addition to the naturally occurring ores there are two forms of enriched ore. The first is titanium slag and is the residue left after extraction of iron. The original ore consists of a mixture of ilmenite and haematite/magnetite which is heated in an electric furnace to approximately 1700°C, when it separates into two layers. The lighter titanium slag layer is run off, quenched and crushed for use as feedstock for pigment manufacture. The two sources are Quebec, Canada, where the TiO_2 content is approx. 72% and Richards Bay, South Africa where the slag comprises 85% TiO_2.

The second enriched ore is beneficiated ilmenite in which the natural ore has been leached at high temperature with either sulphuric acid or hydrochloric acid to remove iron. The beneficiates generally contain 90–96% TiO_2. Typical analyses for various sources of ores and beneficiates are given by Whitehead.[2]

2.2. Sulphate Process

The sulphate process is extremely complex and is said to employ every classical chemical process except distillation. However, the chemistry is very simple:

$$FeTiO_3 + 2H_2SO_4 \longrightarrow TiOSO_4 + FeSO_4 + 2H_2O$$
$$TiOSO_4 + (n+1)H_2O \longrightarrow TiO_2 \cdot nH_2O + H_2SO_4$$
$$TiO_2 \cdot nH_2O \xrightarrow{\text{Heat}} TiO_2 + nH_2O$$

The different manufacturers use their own variants and Fig. 1 is only illustrative of a typical process.

Ores used are ilmenite having a TiO_2 content of less than 60% (above 60% it is not easily dissolved) or enriched ore such as titanium slag.

The ore is dried and ground, the particle size being controlled by air

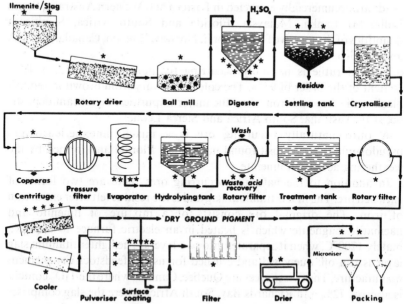

FIG. 1. Diagrammatic representation of the sulphate process. Asterisks denote points of production control tests.

flotation in order to ensure optimum conditions for dissolving. The classified ore and sulphuric acid are added to the digesters which are usually of concrete or steel, lined with lead or acid-resistant tiles. The strength of the acid is usually 85–92% but concentration and quantity are dependent upon analysis of the ore. Digestion may be a batch or continuous operation. The mixture is agitated by compressed air and superheated steam. When the temperature reaches about 160°C a vigorous exothermic reaction occurs and a porous cake is obtained containing titanium, ferrous and ferric sulphates. This cake is extracted with water or dilute (waste) acid. Since iron must be present in the ferrous state for its subsequent complete removal, ferric iron is reduced by addition of scrap iron at a temperature below 70°C. Reduction is continued until some trivalent titanium is formed. The liquor is clarified by sedimentation, aided by the addition of flocculating agents, to remove insoluble residues such as silica, zircon and unreacted ore. The supernatant solution is cooled to about 10°C when most of the iron is crystallised as copperas ($FeSO_4.7H_2O$) and removed by centrifuging. The copperas may be utilised in water and sewage treatment, and in agricultural applications. The crystallisation may be omitted if an ore of sufficiently low iron content, such as titanium slag, has been used. The last traces of sludge are removed by filtration and the solution is concentrated. If ilmenite was the starting material a solution of specific gravity 1·67 at 25°C will contain about 230 g/litre of TiO_2.

Hydrous titanium dioxide is precipitated by hydrolysis. This stage is probably the most critical part of the whole process and it is essential that no ferric iron is present. Precipitation is achieved by leaching for several hours and conditions must be rigidly controlled so that the precipitate can be easily filtered and washed and also produce crystals of the correct type and size when subsequently calcined. Although a precipitate would be obtained by simply boiling the solution, in practice, seeds known as nuclei are added to ensure the correct crystal formation. Nuclei are prepared from pure titanium tetrachloride solution by partial neutralisation and heat treatment and may be produced either in the anatase or rutile form, depending on the method of preparation. The quality of the precipitate depends upon the composition of the main solution, the quality of the added nuclei and duration of hydrolysis. The yield can be increased by diluting the solution when precipitation is nearing completion but quality can be impaired by excessive dilution. The precipitated titanium dioxide is always in the anatase form, regardless of whether the nuclei were anatase or rutile. However, during the subsequent calcination when rutile nuclei are present, the precipitated anatase converts entirely to rutile.

The precipitate is filtered and washed with water, the acid filtrate being recovered and recycled. The precipitated pulp is leached under reducing conditions to remove any last traces of iron, which would have a deleterious effect on colour. Minor additions are made of compounds such as zinc oxide, alumina, sodium and potassium salts, phosphate and antimony oxide, which help to control crystallite growth during calcination.[3,4]

Calcination is carried out in internally fired, inclined rotary kilns through which the pulp moves slowly under gravity. The temperature at the point of entry is about 350°C and, at discharge, 850–1000°C, with retention times of 12–16 h. As it progresses through the kiln the wet pulp is first dried, then strongly adsorbed water and sulphur trioxide are driven off. The acidic gases are washed to convert the sulphur trioxide to sulphuric acid, which is recycled to earlier stages of the process. Crystallite growth and, when relevant, conversion to rutile only occur in the last metre or so of the kiln and careful control of temperature is essential at this stage.[5] In the absence of nuclei, anatase crystals would be formed first and then converted to rutile by prolonged heating at very high temperatures. Addition of rutile nuclei and rutilising catalysts such as zinc and aluminium compounds allows rutilisation to occur more rapidly at lower temperatures, thus preventing excessive crystal growth, sintering of crystals and discolouration, in addition to saving energy.

The pigment is cooled at a controlled rate, as too rapid cooling would result in the formation of trivalent titanium which would adversely affect the colour (greying). The unrefined pigment may then be dry-milled and sold as an untreated pigment, but most production is surface-treated to improve durability, dispersibility and optical properties. This process is discussed later in the chapter.

2.3. Chloride Process
The chloride process depends upon the following chemical reactions.

$$TiO_2 \text{ (impure)} + 2\,Cl_2 + C \longrightarrow TiCl_4 + CO_2$$

$$TiCl_4 + O_2 \longrightarrow TiO_2 \text{ (pure)} + 2\,Cl_2$$

The high cost of chlorine necessitates the use of ores which are rich in titanium dioxide and, consequently, mineral rutile or beneficiated ilmenite is generally used. Where low-cost chlorine is available, lower-grade feedstocks such as leucoxine or slag may be used, but this produces metal chloride by-products which have to be disposed of.

Figure 2 is a schematic diagram representing the chloride process. The dry ore is fed into a chlorinator in which it forms a bed, fluidised by an air

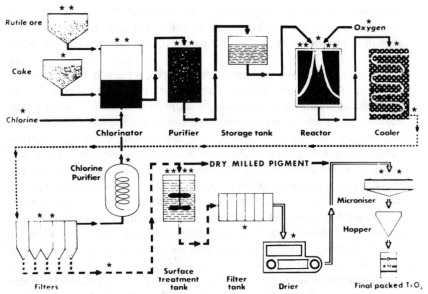

FIG. 2. Diagrammatic representation of the chloride process. Asterisks denote
points of production control tests.

stream. Heat is applied until a temperature of approximately 650°C is
attained. Crushed coke is fed in on top of the ore, where it ignites, increasing
the temperature and enabling the initial heat source to be removed. When
the required temperature (typically 900°C) is attained, the air stream is
replaced by chlorine and the reaction to form titanium tetrachloride
vapour occurs. As the reaction proceeds the chlorinator is continuously
charged with ore/coke mixture to maintain the bed height constant. The
reaction is highly exothermic and it is necessary to cool the reacting mass to
maintain a constant temperature. One method is by controlled addition of
liquid titanium tetrachloride.

The gas stream leaving the chlorinator contains gaseous titanium
tetrachloride, vapour impurities and dust particles. Cooling to about 200°C
in a spray vessel results in most of the impurities condensing as solids on to
the dust particles, and settling to the bottom of the vessel whence they are
continuously removed. After further cooling and cleaning by scrubbing, the
gas is condensed by passing successively through water- and brine-cooled
heat exchangers. The crude titanium tetrachloride is purified by distillation
prior to oxidation to titanium dioxide.

For successful oxidation it is necessary to mix the gases at a suitable
temperature—typically 1200°C—and to provide suitable nuclei on which

the pigment particles may form. The reaction time is measured in milliseconds and the product must be chilled rapidly to prevent excessive growth of pigment particles and formation of massive deposits on reactor walls. The heat evolved when titanium tetrachloride is burned in oxygen is not sufficient to raise and maintain the temperature at the required level. Consequently, additional heat must be supplied. One method is to burn an auxiliary gas such as carbon monoxide and to use the flame to provide a mixture of hot carbon dioxide and aluminium chloride, which is subsequently brought into contact with oxygen. Burners may be used in which titanium tetrachloride, oxygen and auxiliary fuel are fed through separate parts; the design of such burners is very complicated and there are many variants. The nozzles must be kept sufficiently cool to prevent growth of massive titanium dioxide deposits on them, which would impair their operating characteristics. A second method is to heat the oxygen in an electric discharge and introduce it into the reaction system by a plasma torch. A further method is to pass the reacting gases through a hot fluidised bed of inert material such as silica; in this case the titanium dioxide particles are carried out with the gases. A problem with this method is the retention of titanium dioxide as a solid accretion on the particles of the fluidised bed.

Nucleation of reacting gases is necessary to promote the formation of pigmentary particles and may be effected either by introducing a small concentration of water vapour into the oxygen stream or by the combination of hydrocarbons in the 'auxiliary gas' or 'flame' methods. Alternatively, anhydrous $AlCl_3$, added to the tetrachloride feed, is oxidised to Al_2O_3 which provides centres for the growth of titanium dioxide pigment particles.

Irrespective of the type of reactor, the gases carrying the fine pigment must be cooled rapidly to below reaction temperature to prevent further growth. This cooling may be done by using either a mixture of spent gases recycled from a later stage, or liquid chlorine. The pigment is separated by various methods, e.g. cyclone bag filters, Cottrell precipitation, and the chlorine is recovered, purified, compressed and condensed to a liquid and returned to storage.

The pigment is liable to contain adsorbed chlorine which must be removed. The pigment is neutralised by washing and heating or by treatment with steam which may be followed by treatment with air containing steam. The pigment may require milling to break up aggregates, and although it may be used as untreated pigment, usually it is coated as with pigments produced by the sulphate process.

2.4. Sulphate Process versus Chloride Process

Although current sulphate pigment production is almost double that of the chloride, there has been very little change in its level since the mid 1970s; by contrast, chloride capacity has almost trebled, and there is much debate about which will predominate in future. The advantages claimed for the chloride process are improved colour, more efficient processing and reduced environmental consequences. The principal technical advantage is that the purification is carried out by distillation of titanium tetrachloride and hence almost all the impurities can be removed. In the sulphate process purification is via a series of precipitation, filtration and washing steps and removal of the last traces of impurities is not economic, although these are present at only the parts per million level. Consequently, the chloride pigments generally have a higher brightness than the sulphate pigments. However, when the pigments are incorporated into the end-products such as paints, inks, plastics, rubber and paper, these differences in brightness are frequently not significant and when products are toned to give a bluer white or to improve the contrast ratio, the differences in brightness are eliminated altogether. A disadvantage is that, for reasons not yet understood, pigments prepared by the chloride process are more abrasive than their sulphate equivalents and therefore sulphate pigments are preferred in applications where wear may be a problem, e.g. gravure printing inks.

A process advantage is that the chloride route is continuous and more compact and, when high-grade ores are used, almost all the chlorine can be recycled. However, technologically it is more difficult and there are severe high-temperature and corrosion problems. Also, there are toxicity hazards due to the presence of large quantities of chlorine and titanium tetrachloride, and the chlorinated impurities are more toxic than their sulphate counterparts. In addition, although the manufacture of anatase pigments by the chloride process is not impossible, anatase pigments of high purity have not yet been commercially produced; whilst this is not of much importance in the paints and inks industries, considerable quantities are used in textiles, rubber and paper.

Environmentally the chloride process is more attractive. When titanium-rich ores, such as the mineral rutile, are used the amount of impurities is much less and the chlorine can be recycled, whereas in the sulphate process, large amounts of ferrous sulphate and dilute sulphuric acid are produced. Whilst some ferrous sulphate may be used in water and sewage treatment, for the production of iron oxide pigments and in agricultural applications, there is more than sufficient to satisfy these requirements. However, currently there is insufficient mineral rutile available and the use of enriched ores such as beneficiated ilmenite does not eliminate the waste acid problem but transfers it from the pigment manufacturer to the ore producer. When lower-grade ores such as leucoxine are used, significant quantities of waste chlorides are produced and whilst currently these are being disposed of by deep-well injection, this procedure may not be' environmentally acceptable in future.

The long-term expansion of each process will probably depend upon the availability and cost of feedstocks, relative costs of chlorine and sulphuric acid, capital cost of installing new plant—chloride being much more expensive—and cost of treating the effluents.

2.5. Coating

The properties of titanium dioxide pigments can be substantially improved by coating the surface with hydrated inorganic oxides. The original purpose of coating was to reduce the chalking tendency and yellowing that occurred with certain types of paints when exposed outdoors, and also to prevent the fading when mixed with coloured pigments. However, it was subsequently found that the correct choice of coatings could also improve the dispersibility and optical properties of the pigments.

The coating agent must be a colourless compound; various hydrated forms of alumina and of silica are commonly used but others, such as titania and zirconia, are used for particular purposes.

The untreated pigment, whether produced by the sulphate or chloride process, is dispersed in water by ball-mill or sand-mill, to ensure that oversized particles consisting of sintered aggregates are eliminated. Sodium silicate or an organic dispersing agent may be used to facilitate dispersion and the slurry may be allowed to settle, thus removing any remaining oversized particles. The coating reagents are added to the dispersion, which is continuously agitated.

Careful control of pH and temperature are essential to ensure that the coatings are precipitated on to the pigment surface in the correct sequence and in the desired form.

The inorganic oxide coating process consists of precipitating the hydrous oxide on to the surface of the pigment particles and may be represented simply by

$$\text{Silicate} + \text{acid} \longrightarrow \text{hydrated silica}$$

$$\text{Aluminate} + \text{acid} \longrightarrow \text{hydrated alumina}$$

$$\text{Aluminium salt} + \text{alkali} \longrightarrow \text{hydrated alumina}$$

In practice, the process is much more complex. When mixed coatings are applied, conditions may be chosen so that one or other oxide is completely precipitated before coating with the second, or both may be precipitated simultaneously. It is also essential to ensure that the coatings adhere firmly to the pigment surface rather than being present as an admixture. A typical difference in durability of a paint film containing a coated pigment or a blend of pigment plus alumina was given by Evans;[6] durability was measured by weight loss (Table 1).

TABLE 1

Pigment	Wt loss (mg/cm^2)
TiO_2 with alumina coating	0·50
TiO_2 blended with alumina	1·18

A further complication is that hydrous alumina can exist in various forms, depending upon pH and temperature, ionic strength and type of reagent in the coating tank. These various alumina forms will have different effects on the pigment properties. Thus, conditions such as pH and temperature must be rigidly controlled, and agitation should be very good to ensure uniform conditions throughout the tank, with no localised areas of differing pH. There are numerous references to coating compositions and techniques, in the patent literature, all drawing attention to the precision required.[7-14]

The quantities of coatings applied are usually of the order of 3–5% but values as high as 15% may be used for pigments intended for matt coatings.

The pigments are dried to a specified moisture content and then given a final milling, generally in some form of fluid energy mill such as a microniser. The ratio of steam to pigment, rate of feed and design of the mill are controlling factors in the efficiency of milling. Organic compounds are commonly added to improve milling efficiency: these are added either prior to, or at, the microniser. A variety of organic additives are used, mainly polyols or amines for pigments intended for paint and printing inks, or silicone derivatives for plastics applications. Polyols and amines are commonly accepted as acting as micronising aids to break down agglomerates formed during the drying operation, thereby increasing the ease of dispersion when subsequently incorporated into the paint or ink. The compounds do not act as dispersing aids because addition of the organic compound to a non-organic treated pigment at the paint dispersion stage does not give the same degree of dispersion as a pigment which was given the same organic treatment at the microniser.

The effect of different coatings on various properties will be discussed under the separate headings.

3. PROPERTIES OF TITANIUM DIOXIDE

Pure titanium dioxide (TiO_2) is a colourless, crystalline solid. As with other dioxides of d block elements in Group VI of the periodic table, it is stable, non-volatile, insoluble and is rendered refractory by ignition. It is amphoteric but has more acidic than basic properties.

Titanium dioxide exists in three crystalline forms, rutile, anatase and brookite, but only rutile and anatase have commercial applications.

TABLE 2

Property	Rutile	Anatase
Crystal structure	Tetragonal	Tetragonal
Density	4·23	3·9
Refractive index	2·74	2·50
Hardness (Mohs' scale)	7–7·5	5·5–6·0

Table 2 lists some relevant physical properties and Fig. 3 shows the crystal structures. Although rutile and anatase are chemically identical, and both crystallise tetragonally, they can be distinguished by X-ray diffraction.

The rutile crystal is the more compact and this results in important differences, notably higher refractive index, resulting in better opacifying power, higher density, and greater stability. It melts at 1825°C whereas anatase has no specific melting point, being irreversibly transformed to rutile before a melting point is reached.

Rutile pigments are generally preferred to anatase because of their greater opacifying power and stability, although anatase pigments may be used in some specialised applications, for example:

(a) Paints where self-cleaning is desired—dirt being continuously removed due to the chalking action.

(b) Specialist inks where the least possible abrasivity is required (note differences in hardness in Table 2), e.g. certain gravure inks and inks for use on aluminium foil which is subsequently cut. It has been

(a)

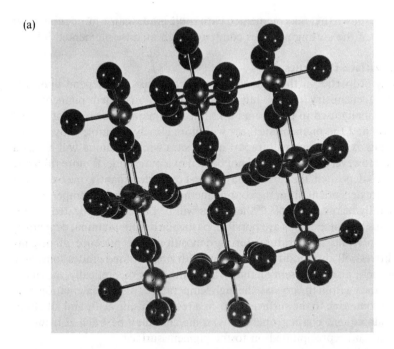

(b)

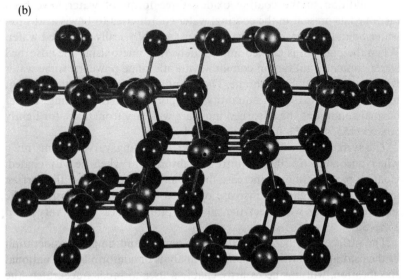

FIG. 3. Crystal structures: (a) rutile; (b) anatase.

shown that inks pigmented with rutile grades cause more rapid wear of the cutting punches compared with anatase-pigmented inks.

3.1. Surface Properties[15-18]

Many properties of the pigmented paints and inks will depend upon the surface chemistry of the pigments. Although uncoated pigments are occasionally used in paints and inks, the vast majority consists of coated grades and the pigment surfaces will more closely resemble those of the coating oxides. Silica will impart acidic sites whilst alumina will act as a base. However the surface will not be that of a pure oxide. If more than one oxide is deposited, the surface may consist of predominantly one oxide or a mixture dependent upon the coating method. As the drying temperature is generally below 150°/160°C the oxides will be present in hydrated forms; various hydrated forms are possible, particularly with alumina, depending upon pH and temperature during deposition. The presence of organic additives will also modify the surface. With zinc-rutilised grades some zinc oxide may also be present on the surface. Although, theoretically, the zinc is contained within the crystal, the high temperatures used in calcination may force some zinc to the surface. Also, as a result of using acids and alkalis in the subsequent coating operations, some zinc may be leached from the crystal and reprecipitated on to the pigment surface.

In addition to the coating oxides, three forms of water have been identified as present in the coating; water coordinated to Lewis acid sites; water bound to surface hydroxyl groups; and physically adsorbed water. When the pigment is packed immediately after micronising the adsorbed water content is nil but, in common with other fine powders, some water will be adsorbed during storage. The quantity will depend on the ambient humidity and also on the quantity and porosity of the coating. Under normal conditions the adsorbed moisture will vary from 0·5% for lightly coated grades to 2% for heavily coated grades.

The surface area will also depend upon the coating, varying from $7m^2/g$ when uncoated, to $30m^2/g$ (nitrogen absorption) when heavily coated. Coating particles should increase their diameter and decrease the surface area per unit weight; the increase observed indicates the porosity of the surface. Porosity will affect the oil absorption values which vary from 15 to 35.

The surfaces are therefore extremely complex and simple consideration of the coating quantities, obtained by analysis or conformity to a national specification, will not necessarily enable a user to predict accurately the results which will be obtained in any particular application.[19]

4. OPACITY

A full discussion on opacity arising from the scattering of light would, itself, require a full chapter. For the various mathematical treatments the reader is referred to refs 20–33. It is intended here only to outline the various factors that affect the opacity of practical paints and inks pigmented with titanium dioxide.

Although, in the massive state, titanium dioxide is colourless, it is an extremely effective white pigment when finely divided for it absorbs almost no light in the visible region of the spectrum.

Light incident on a large crystal of titanium dioxide would undergo some reflection at the interface and refraction within the crystal but sufficient would be transmitted to render the crystal transparent. In its finely divided pigment state however, incident light is scattered, giving an opaque appearance.

The two factors governing the degree of scattering by a small particle are its refractive index or, to be exact, the ratio of the refractive indices of the particle and the surrounding medium, and the size of the particle.

An approximate indication of the opacifying power can be obtained by using the simplified form of Fresnel's reflectance equation:

$$F = \frac{(n_1 - n_2)^2}{(n_1 + n_2)^2}$$

where F is the reflectivity, n_1 is the refractive index of pigment and n_2 is the refractive index of medium.

Table 3 lists the refractive indices of some common white pigments and the reflectivity values obtained by Fresnel's equation, assuming a refractive index of 1·50 for the medium.

TABLE 3

Pigment	Refractive index	Reflectivity (%)	Tinting strength (Reynolds)
Titanium dioxide (rutile)	2·70	8·2	1 850
Titanium dioxide (anatase)	2·55	6·7	1 350
Zinc sulphide	2·37	5·1	900
Antimony oxide	2·29	4·3	400
Zinc oxide	2·02	2·2	200
White lead	2·00	2·0	100
Lithopone (30%)	1·84	1·0	300
China clay	1·57	0·05	< 100

Tinting strength or, more correctly for white pigments, lightening power, indicates the scattering power of a white pigment. It is the amount of coloured pigment, usually blue or black, required to give a standard depth of shade with a given white pigment, relative to white lead having a value of 100.

Table 3 shows that rutile titanium dioxide has a unique superiority as an opacifying agent, a consequence of it having the highest refractive index, the greatest reflectivity and highest tinting strength (lightening power).

Refractive index is not actually a unique value but is dependent upon the wavelength of the incident light, increasing with decreasing wavelength, i.e. blue light will be reflected more than red.

4.1. Particle Size—Scattering Theories

An equation for the intensity of light scattered by particles was first given by Rayleigh[20] in 1871.

The intensity of scattered light is proportional to n^4, and to $1/\lambda^4$, where n is the ratio of refractive index of particle to that of medium, and λ is the wavelength of incident light. The inverse relationship between the intensity of scattered light and λ^4 explains the blue light of the sky, as the shorter the wavelength (i.e. the bluer the light), the greater the scattering by small particles in the Earth's atmosphere. The expression is true only for particles having a refractive index slightly higher than that of the medium and whose dimensions are much smaller than the wavelength of the incident light.

The first mathematical theory for larger particles with a refractive index significantly greater than that of the medium was put forward by Mie[21] in 1908. His theory assumed that the particles were homogeneous and spherical and that the spacing between particles was sufficiently large to exclude any interaction (over $50d$ where d is the diameter of the particle). In practice, pigment particles are not spherical and the spacing required means that the pigment volume concentration (PVC) must be less than about 2%, far removed from that for a practical paint. Nevertheless, many workers have used his theory as a basis for calculating the optimum particle size for maximum light scattering, modifying the theory to account for multiple scattering. The resulting expressions are extremely complex, requiring the aid of a computer for solution.

In simple terms, scattering is dependent upon two parameters:

(1) The ratio of the refractive indices of the particle and surrounding medium (n).

(2) The ratio of the particle size (d) to the wavelength (λ) of the incident radiation.

Thus for any given wavelength there is an optimum particle size for maximum scattering.

The Mie theory also shows that for a given value of n there is an optimum value of d/λ. The larger the refractive index n, the smaller the optimum particle diameter.

Kubelka and Munk [22] in 1931 considered the film as a single entity rather than being comprised of separate scattering centres. They derived a series of equations relating to the scattering (S) and absorption (K) coefficients of the pigmented film. For a white pigment K is very low (zero in a perfect white system) and only the scattering coefficient need be considered.

Their approach was phenomenological and the coefficients cannot be calculated from basic theory. However, many workers have derived experimental values for the scattering coefficients and related them to particle size, i.e. combining the Mie and Kubelka–Munk theories.

The basic Kubelka–Munk theory neglects reflection at the paint film/air and paint film/substrate interfaces and necessary modifications have been made by many workers, e.g. Ross,[31] and Tunstall and Dowling.[32]

4.2. Optimum Particle Size

It is obviously very important that the pigment particles in the film should have the maximum scattering power and it can be shown that for a given wavelength there is an optimum particle size (as distinct from crystal size) for maximum scattering; a typical curve is shown in Fig. 4. The optimum particle size will depend upon the wavelength of light: the longer the wavelength, the larger the optimum particle size.

Values for optimum particle size obtained by different workers vary but

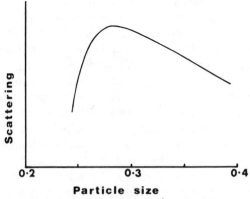

FIG. 4.   Variation of scattering power (S) with particle size.

the discrepancies can be partly explained by the different methods used for determining the particle size, differences in size distribution, pigment loading, wavelength used and film thicknesses. In some older literature there also appears to be some confusion between particle size and crystal size. The shapes of the curves are generally as shown, i.e. a particle size slightly lower than optimum has a much more dramatic effect on scattering power than slightly higher sizes.

The particle size and size distribution will obviously depend upon the mean crystal size and crystal size distribution but will also be significantly affected by the state of dispersion. The crystal size and size distribution are determined during the manufacture of the pigment but the state of dispersion is governed by the paint or ink manufacturer. A narrow particle size distribution is the most effective.

4.2.1. Optimum crystal size
As previously stated, opacity will ultimately depend upon crystal size and crystal size distribution and titanium dioxide pigments with a single crystal size cannot be manufactured; the minimum attainable standard deviation is of the order of 1·25–1·30. The effective particle size is also dependent upon pigment loading, as the greater the loading the closer the packing of the particles until eventually each particle no longer acts as a single scattering entity. In deciding on the optimum mean crystal size the pigment manufacturer has to take these factors into account and also the degree of dispersion likely to be attained by the pigment user.

4.2.2. Effect of pigment loading
In Fig. 5 the scattering coefficient S (equivalent to opacity for a white paint) has been plotted against pigment volume concentration (PVC). Scattering increases linearly with pigment loading up to 10–12% PVC—the exact value depends on the particle size—when the particle separation becomes approximately equal to the wavelength of the incident light. Above this level the relationship will no longer be linear and although the total scattering will increase because of increase in the number of particles, the scattering efficiency of the particles will decrease because of interference or crowding. This is the reason why many workers stipulate that the minimum distance of separation for maximum scattering efficiency should be not less than the wavelength of the incident light. The loss in scattering efficiency increases with further pigment loading and eventually the interference or crowding effect becomes so great that a loss in total scattering occurs. The PVC at which this occurs is usually in the 27–30% range but this value

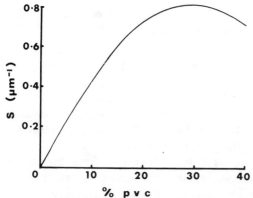

FIG. 5. Variation of scattering power (S) with PVC.

decreases as particle size and film thickness decrease. In extremely thin films, e.g. printing inks, the maxima may occur at values as low as 23%. The effect of pigment loading on scattering efficiency E_f is plotted in Fig. 6 (as dots) using data contained in Fig. 5, where E_f is calculated as follows:

$$E_f = \frac{S/PVC}{\underset{PVC \to 0}{Lt}\ (S/PVC)}$$

Various investigators have quantified the effect of pigment crowding, e.g. Ross,[31] and Tunstall and Hird.[33]

Tunstall and Hird derived an expression which showed that the scattering efficiency, E_f, depends upon the pigment volume fraction f

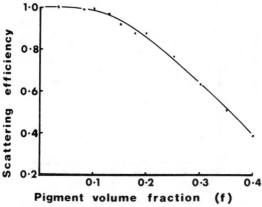

FIG. 6. Effect of pigment loading on scattering efficiency (E_f).

(100f = PVC), the mean surface-to-surface separation distance c and the wavelength of the incident light λ.

$$E_f = 1 - \frac{1}{f}\exp\left(-13\cdot15\frac{c}{\lambda} - 1\right)$$

where $c = d(\sqrt[3]{0\cdot49/f} - 1)$, d = mean particle size.

As pigment loading f increases, the interparticle distance decreases and scattering efficiency decreases. Also as f increases greater scattering will be obtained if a larger particle size is used.

Values of E_f have been calculated from the above equation and superimposed (solid line) on the experimental data in Fig. 6 for $\lambda = 0\cdot55\ \mu m$ and $d = 0\cdot275$. The value for d was determined experimentally using a light-scattering technique.[34] Agreement between the experimental data and calculated values is very good.

4.2.3. Mean crystal size and particle size

The Tunstall and Hird equation can be used to plot the relationship between crystal size, particle size and PVC to give maximum opacity (scattering power), as shown in Fig. 7. A standard deviation of 1·30 for crystal size was used and a log–normal distribution for both crystal size and particle size was assumed. The single crystal fraction (F_1) in the dry film was determined by electron microscopy and this is taken as an indication of mean particle size. As F_1 decreases, mean particle size increases.

As would be expected from the equation, for a given degree of dispersion

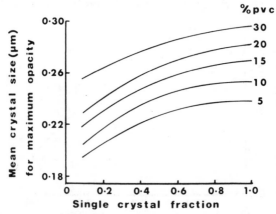

FIG. 7.   Theoretical relationship between crystal size and single crystal fraction for maximum opacity at different pigment loadings.

(i.e. constant F_1) the mean crystal size for maximum dispersion increases with PVC.

Many investigators have shown that in paint films where the state of dispersion would be considered as good, the number of particles present as single crystals is only of the order of 20%, i.e. single crystal fractions of 0·2. Thus, for high-gloss paints pigmented at PVCs of 15–20%, the optimum mean crystal size is between 0·22 and 0·24 μm.

Figure 7 also shows that at higher PVCs, such as may be used in metal decorative paints and printing inks, the optimum mean crystal size should be increased. However, as the crystal size is increased the wavelength which is preferentially scattered is also increased, i.e. less blue and more red light will be scattered and the pigment will have a browner undertone. Generally, blue undertones are preferred, even at the expense of a slight loss in opacity. In addition an increase in crystal size will also reduce gloss potential in thin films.[35]

4.2.4. Effect of dispersion on scattering

Modern titanium dioxide pigments, particularly those which have been organically treated at the micronising stage, disperse extremely rapidly in most paint media and the more important factor is maintaining that dispersion, i.e. preventing flocculation (discussed further on).

Table 4 gives the opacity, at a spreading rate of 20 m²/litre, fineness of grind and particle size data (in the wet paint) which include the standard deviation as a function of milling time (ball-mill) for a gloss-paint-based on a long-oil soya-modified alkyd pigmented at 17·5% PVC. Maximum opacity was obtained after only 1 h and, indeed, there was only a small difference between that and the hand-stirred sample. This result indicates that the majority of the pigment dispersed easily leaving only a small

TABLE 4

Milling time (h)	Contrast ratio at 20 m²/litre (%)	Fineness of grind (μm)	Particle size (μm)	
			Mean	Standard deviation
Hand-stirred	94·3	> 50	0·32	1·53
0·25	94·6	20–42	0·30	1·48
1	94·8	10–18	0·29	1·48
4	94·8	6–14	0·29	1·47
20	94·8	5–14	0·28	1·47
70	94·9	5–12	0·28	1·47

proportion of oversize particles. However these would result in 'bitty' film, as it requires only 0·1% of the pigment to be present as oversize particles to give an unacceptable fineness of grind rating. Thus, whilst 4 h was required to give an acceptable appearance, milling for only 1 h resulted in maximum opacity. Extending the milling to 70 h had no effect on opacity. However, in Fig. 4, the rapid decrease in scattering at particle sizes below the optimum value suggests that overmilling could result in a decrease in opacity. Figure 7 also shows that an increase in the single fraction size (i.e. decrease in particle size) requires an increase in mean crystal size in order to maintain maximum opacity, which obviously will not occur in practice. This apparent contradiction regarding the effect of overmilling can be explained by a small degree of flocculation, which inevitably occurs in the wet paint and generally increases on drying, which is sufficient to mask any differences due to an increase in the single crystal fraction. Slight shear will generally increase the scattering of wet paints whilst electron microscopy shows that slight flocculation is the rule, rather than an exception.[36]

Factors affecting dispersion and flocculation will be discussed further on.

4.3. Effect of PVC on Opacity for Different Grades of Pigment

Figure 5 shows the variation in scattering power for PVCs of up to 40%. This curve is for a pigment having a TiO_2 content of 94·5% and intended for use in gloss paints. As the pigment loading is increased further, the scattering power continues to decrease until the critical PVC (CPVC) is reached, where there is just sufficient binder to satisfy the absorptive properties of the pigment. As the CPVC is exceeded, air will become entrapped in the paint film and because the refractive index of air is 1·0, the effective refractive index of the binder (now medium + air) will decrease and a rapid increase in scattering will occur as shown in Fig. 8 (pigment A). This sudden increase is often referred to as dry hiding.

With heavily coated grades of titanium dioxide (pigment B), intended for use in matt paints, the TiO_2 content is lower (80%). As the scattering in fully bound systems is dependent upon TiO_2 content, up to about 40% PVC, pigment A has the better opacity because of its higher TiO_2 content: the inorganic oxide coatings on the titanium dioxide crystals have refractive indices of the same order as binders i.e. 1·5, and therefore do not contribute to scattering in fully bound systems.

However, because of the amount and porous nature of the coating on pigment B (surface area 30 m²/g), the critical PVC will be at a lower value than that of pigment A (surface area 13 m²/g) and, above about 40% PVC, pigment B will give better scattering.

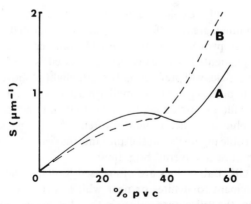

FIG. 8.   Variation of scattering power(s) with PVC for different types of pigment.

Note, however, that with emulsion paints there is a tendency for air to be entrapped in the film at PVCs below the critical value, and this will partially compensate for the loss in scattering due to pigment crowding. Accordingly, the drop in scattering above 30% PVC for pigment A, and the point of inflexion with pigment B, will not be as marked compared with an air-drying alkyd paint.

5.   COLOUR

5.1. Whiteness

The reflectance curves for the two crystal forms are given in Fig. 9. The slight absorption at the blue end of the spectrum accounts for their slightly cream tone compared with a perfect white such as magnesium oxide. The cream tone is more noticeable with rutile pigments. The two factors that effect whiteness are purity and particle size. In order to ensure good

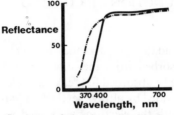

Fig. 9.   Spectral reflectance of titanium dioxide for paint films of 20% PVC:
·—·—, anatase; ——, rutile.

whiteness removal of the coloured impurities present in the ore is essential as they have a much greater effect if they are incorporated into the crystal rather than when present as admixtures. The difference between chloride and sulphate pigments has already been discussed in Section 2.4.

As smaller particles will preferentially scatter light of shorter wavelength, appearing more blue, pigments of smaller particle size, and smaller crystal size if all other factors are equal, will tend to counteract the yellow tone. Toning with blue or violet tinters is often used to mask the inherent yellowness of rutile pigments but the amounts that can be added are limited as they also reduce the overall brightness.

Whilst the CIE Lab. values are often used to define the colour, a very useful measurement for whites or near whites is the Colour Index (CI), calculated from the reflectance values in the blue, green and red region of the spectrum:

$$CI = \frac{R - B}{G} \times 100$$

For a pure white, R, B and G would all equal 100 and therefore CI would equal 0. A negative value denotes a blue tone and a positive value a cream tone.

5.2. Undertone

This is significant in colour matching and is generally assessed by mixing with a specified quantity of a black pigment. The tone of the resultant paste or paint is compared with a standard and rated as bluer or browner. Instrumental measurements may also be used.

The tone is dependent upon the particle size: the smaller the particles, the bluer the undertone. Thus, whilst it is dependent upon the mean crystal size, the undertone can also be affected by the degree of dispersion.

6. FUNCTION OF COATINGS

6.1. Durability[37-42]

The photochemical degradation of a binder is primarily an oxidation process with oxygen absorbed from the atmosphere. For oxidation to proceed, energy must also be absorbed to break the molecular bonds in the binder. This energy is provided by ultraviolet (UV) radiation in the form of photons which collide with the electrons associated with the molecular bond. Every chemical bond requires a minimum energy level for disruption,

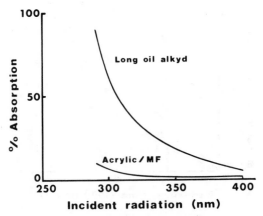

FIG. 10.   Absorption of radiation by unpigmented films (20 μm thickness).

which accounts for different binders exhibiting different rates of breakdown.

The rate of photochemical breakdown is dependent upon:

(1) The energy provided, i.e. the wavelength of the incident radiation— the shorter the wavelength, the greater the energy.
(2) The intensity of the radiation.
(3) The absorption characteristics of the binder.

Figure 10 illustrates the difference in the absorption at different wavelengths for a long-oil air-drying alkyd and an acrylic/melamine– formaldehyde (MF) stoving system, which is in agreement with the observed faster degradation of the long-oil alkyd on exterior exposure.

In the case of a coating pigmented with titanium dioxide, most of the incident ultraviolet radiation is absorbed by the titanium dioxide. Thus the presence of rutile pigment in a medium should reduce the potential rate of degradation of that medium. Unfortunately, titanium pigments are also photoactive and as a result can catalyse the breakdown of the binder (photocatalytic degradation). Accordingly, the presence of titanium dioxide pigment can accelerate or retard the degradation, depending on whether the photoactivity of the pigment is greater, or less, than the rate of breakdown of the binder alone.

Figures 11(a) and (b) show the weight loss of alkyd and polyurethane (polyester/aliphatic isocyanate) paints pigmented with increasing quantities of titanium dioxide pigment. Increasing the quantity of pigment 'protects' the alkyd but accelerates the degradation of the polyurethane.

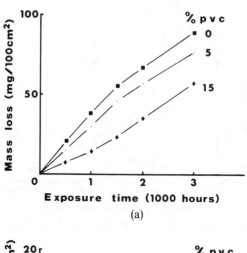

(a)

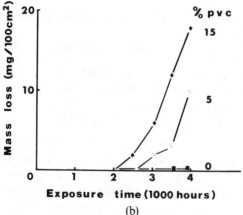

(b)

Fig. 11.   Effect of PVC on mass loss: (a) moderately durable resin (long-oil alkyd);
(b) highly durable resin (polyurethane).

The photocatalytic degradation is due to the formation of hydroxy and perhydroxy radicals, both of which are highly reactive and can initiate breakdown of organic compounds such as resins. The precise mechanism of their formation has been the subject of much debate but a general overall mechanism is as follows.

Titanium dioxide pigments consist of crystals that contain a variety of defects e.g. Ti^{4+} or O^{2-} ions are missing from the lattice or have been substituted by other ions. When a rutile crystal is irradiated, provided that the quantum of energy is in excess of 282 kJ/mol—equivalent to radiation with a wavelength less than 405 nm—electrons are excited from the valence band to the conduction band leaving positive holes in the valence band:

$$TiO_2 \xrightarrow{hv} e + h$$

The excited electrons and holes (excitons) are free to move within the crystal lattice. Some may recombine

$$e + h \longrightarrow \text{thermal energy}$$

but others will reach the surface where they will react to form the hydroxy and perhydroxy radicals.

(1) The positive holes will react with hydroxy groups, present on the surface to form adsorbed hydroxy radicals:

$$h + OH^- \longrightarrow OH^{\cdot}(ads)$$

(2) The electrons react with adsorbed oxygen to form O_2^- radical ions which then react with water to form the perhydroxy radical:

$$e + O_2 \longrightarrow O_2^-(ads.)$$

$$O_2^-(ads.) + H_2O \longrightarrow OH^{\cdot} + HO_2^{\cdot}$$

The above reactions are evidence that both oxygen and moisture are essential for photocatalytic degradation to occur, whereas in photochemical degradation of binders, only oxygen is essential, although water will catalyse the reaction.

The coatings of hydrous oxides (most commonly silica, alumina and zirconia) reduce the photocatalytic effect by acting as electron acceptors and by providing an 'active' area for recombination of the hydroxy radicals.

$$2OH^{\cdot}(ads.) \longrightarrow H_2O + \tfrac{1}{2}O_2$$

The photocatalytic activity can also be reduced by restricting the movement of the excitons through the defective crystal lattice by the judicious introduction of conditioning elements into the lattice. Traditionally, zinc and aluminium are used; they act as electron and hole recombination centres, thereby reducing the number that reaches the surface.

Anatase pigments are much more photoactive than rutile grades and, although coating with inorganic oxides will reduce their photoactivity, coated anatase grades are still not suitable for use in paint intended for exterior exposure.

Figure 12 illustrates the typical differences in durability (as measured by

T. ENTWISTLE

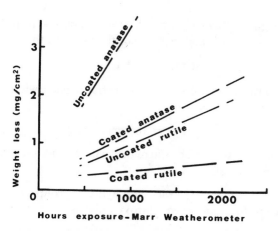

FIG. 12. Effect of different types of pigment on mass loss (long-oil alkyd P:B 1:1).

film weight loss) of a long-oil air-drying alkyd pigmented with uncoated and coated anatase and rutile pigments. The rutile content of the pigmentary crystal is extremely important and great care is essential during manufacture to ensure a high value. The increase in photocatalytic activity with decreasing rutile content in the crystal is much greater than would be expected, e.g. the photoactivity of a pigment of 90% rutilisation is double that of one of 100%.[6]

Reasons for the greater photocatalytic activity of anatase compared with rutile have not been fully explained. The greater absorption of UV radiation by rutile (Fig. 9) and the greater scattering of incident radiation arising from its higher refraction index will ensure better protection of the binder, but will not account for the lower photoactivity of rutile. Contributory factors may be:

(1) The proposed mechanism for the formation of perhydroxy radicals involves the reaction between electrons and adsorbed oxygen and the rate of photoadsorption of oxygen by anatase is double that of rutile.[43]

(2) The inclusion of zinc and aluminium ions as rutilising agents reduces the number of crystal defects and hence will hinder the migration of excitons to the surface and/or provide centres for the recombination of the electrons and holes.

The reader should note that great care is required when comparing the photocatalytic activities of different titanium dioxide pigments using accelerated weathering machines. Egerton and King[38] have shown that, whereas photochemical degradation of binders increases with decreasing wavelength (increasing energy), the photoactivity of titanium dioxide is independent of wavelength below a threshold value of 405 nm. Thus when using fluorescent tubes which emit relatively low-wavelength radiation (typically 280–320 nm) together with condensation to simulate rain and dew, the photocatalytic activity of the pigments will not be affected, although the greater energy will increase the photochemical degradation of the binder. Thus, the ratio of photochemical to photocatalytic degradation will increase with decreasing wavelength and the use of artificial shorter-wavelength radiation will not show the differences, obtained on natural weathering, between grades of titanium dioxide pigments.

6.2. Dispersion
Although the coatings of inorganic oxides were originally applied to reduce the photoactivity of the titanium dioxide pigments, it was quickly realised that suitable coatings could be designed which would also improve the ease of dispersion. However, the design of pigments with both maximum durability and maximum ease of dispersion is not always possible, which partially accounts for the multitude of grades offered by each pigment manufacturer.

Before discussing the function of coatings in dispersion, a brief consideration of what is implied by dispersion will be useful. Whilst pigment manufacturers take great care to ensure that the crystals are of the correct size to impart maximum opacity, the crystals will tend to cohere during the later stages of manufacture and on storage to form larger particles. Three types of particles have been described by Gerstner.[44]

(1) Aggregates—groups of crystals joined at their faces and having a surface area significantly less than the sum of the areas of their constituents.

(2) Agglomerates—groups of crystals and aggregates, which are joined at their edges and corners and having a surface area not markedly different from the sum of the areas of their constituents.

(3) Flocculates—loose assemblies in which the forces between the particles (crystals or aggregates) are very much weaker than in aggregates.

In the case of titanium dioxide pigments the aggregates are mainly formed during calcination (sulphate process) or oxidation (chloride process) although some may be formed during drying after coating; on the other hand, agglomerates and flocculates are mainly formed during drying or storage of the pigment.

Over the past 25 years the number, and size, of aggregates present in titanium dioxide pigments have been very much reduced, by improved milling prior to coating, and also by the use of micronising, particularly when treated with organic additives, as the final stage after drying, immediately prior to packing. These improvements have contributed to the improvement in ease of dispersion.

The process of dispersion can be regarded as consisting of three stages:

(1) Wetting—the replacement of pigment/air interfaces by pigment/ medium interfaces.

(2) Disruption—the breakdown of agglomerates, flocculates and possibly aggregates, although in the case of high-speed impeller mills it is unlikely that the forces generated are sufficient to break down aggregates.

(3) Stabilisation—the maintenance of the disperse state.

Uncoated titanium dioxide pigments have a high tensile strength making them extremely cohesive and resulting in wetting and disruption being very difficult. Coating with inorganic oxides reduces the tensile strength, making wetting and disruption much easier. A method for determining the tensile strength is described by Parfitt.[45]

The combination of coating and improved milling during pigment manufacture has resulted in modern-day titanium dioxide pigments being easy to disperse, as illustrated in the rapid attainment of maximum opacity given in Table 5 (ease of dispersion is dependent on both resin and solvent). The major problem is that of dispersion stabilisation, i.e. prevention of flocculation both during storage and while drying after application of the paint or ink. The function of the pigment coating is to assist in preventing this flocculation, in addition to reducing the degree of cohesion.

In the absence of repulsive forces, the dispersed particles will tend to flocculate under the action of London–van der Waals forces of attraction. These forces become effective when the interparticle distance is less than 1 μm and increase rapidly as that distance decreases.[46] Dispersion stability, or freedom from flocculation, can only be maintained if the pigment particles are separated sufficiently for the repulsive forces to exceed those of attraction. Two mechanisms for stabilising dispersions are possible:

(1) Electrostatic repulsion due to the presence of charges on the particle surface. The repulsion between the counterions (electrical double layer) in the liquid phase keeps the particles separated,

(2) Steric repulsion resulting from the adsorption of polymer chains on to the particle surface.

There is general acceptance that in concentrated systems, such as exist in paints and inks, steric repulsion is mainly responsible for dispersion stabilisation.

The adsorption of polymers on to pigment surfaces has been widely studied but investigation is complicated by the complex nature of the surfaces, a topic discussed in Section 3.1. In addition, the resins used in paints and inks are not pure compounds but vary in composition, molar mass and the number of polar groups. To enable scientific studies to be made, many workers have used pure oxides or have dried pigments, to remove water and hydroxyl groups prior to adsorption, or have used simple compounds as adsorbents. Such studies have been invaluable in increasing our understanding of adsorption of polymers on to pigment surfaces, but care is required in interpreting the results to allow for the complexities of surfaces and adsorbents that are present with commercial pigments and resins.

Experiments with radioactively labelled polymers have shown that adsorption is dynamic, i.e. polymers can be desorbed and re-adsorbed. With uncoated grades equilibrium is rapidly attained, indicating that the adsorption is physical, whereas with coated grades interchange is much slower, in some cases weeks or even months being required to reach equilibrium, i.e. adsorption is chemical.

The acidic silica, basic alumina and the many hydroxyl groups present in their various hydrated forms, and also combined water, will all contribute polar sites. The manner in which polymers are adsorbed on these polar sites determines the degree of dispersion stabilisation. Calculations[47] show that for particles of 0·2 μm diameter, dispersed in hydrocarbon solvent, the minimum separation distance to prevent irreversible flocculation is 10 nm, so that the adsorbed layer must have a minimum effective thickness of 5 nm, i.e. the polymer molecules must be well extended from the pigment surface.

All alkyds are the class of resins most widely used in paints, these have been the subject of many studies. Alkyd molecules will contain typically 1–3 COOH groups, more likely 1–2, distributed randomly along the backbone of the molecule. Other polar groups will be ester, hydroxyl and aromatic groups of the acid, all of which are basic, and these will be much more numerous than the acid groups. With pigments predominantly coated with silica, there will be many contacts between the numerous basic groups of the alkyd molecules and the acidic silica sites so that the polymer molecules

TABLE 5

Pigment	Coating levels		Resin adsorbed (mg/g)	Solvent adsorbed (mg/g)	Relative opacity
	SiO_2	Al_2O_3			
A	—	3	27·6	122·7	100
B	0·5	1·5	23·3	48·9	95
C	3	—	13·1	—	83

will lie flat on the pigment surface. Thus, the adsorbed layer will be compact and not have sufficient extension from the pigment surface to prevent flocculation. If the pigment coating is principally alumina, there will be strong interaction between the basic alumina sites and the acid groups of the alkyd. As these are small in number there will probably be only one or two points of attachment. As the pigment surface will contain few acidic sites to attract the basic groups present in the alkyd molecule, the polymer molecules will extend from the pigment surface and prevent flocculation. Table 5 gives details of the amount of resin adsorbed by pigments of different silica–alumina coatings[48,49] and their relative hiding power—film thickness required to give a standard contrast ratio.

Opacity, which is related to freedom from flocculation, is dependent upon the adsorption of both resin and solvent, and an essential aspect of the theory of steric repulsion is that the adsorbed polymer molecules should be well solvated.

Adsorption of the non-polar solvent, used in these experiments, on the pigment's polar surface is unlikely and, accordingly, the solvent must be entrapped in the loops of the polymer molecules. The configuration of the polymer molecules will determine the amount of space available in the adsorbed layer for trapped solvent; the larger the loops, the greater will be the amount of solvent. Solvent adsorption is thus a consequence of resin adsorption and not a competing process. Where the resin layer is compact (pigment C) there will be no loops and solvent adsorption will be minimal. Franklin[48,50] showed that opacity—i.e. dispersion or freedom from flocculation—increased with increasing alumina content up to approximately 2%, indicating that sufficient alumina is necessary to provide the necessary sites for attachment of carboxyl groups. He also found that zeta potential became more positive with increasing alumina content but the actual values were dependent upon the resin system; solvent is also a contributory factor to zeta potential. In some cases the potential increased from a negative value towards zero. As electrostatic repulsion is proportional to the square of the charge, hiding power and dispersion should be independent of the sign of the charge. As opacity increased with decreasing negative charge, electrostatic repulsion cannot be the controlling factor. Also with all the resins used, increasing opacity was obtained with zeta potentials less than 10 mV, which is insufficient to give rise to stable dispersions.[51]

If extension of the polymer molecules from the pigment surface and the size of the polymer loops and coils is a controlling factor in steric stabilisation, it is reasonable to expect that opacity will increase with molar mass of the polymer and this has indeed been found to be true.[48-50] Moreover, Goldsbrough and Peacock[49] found that although opacity increased with molar mass, the zeta potential decreased, providing further evidence that stabilisation was not attributable to electrostatic repulsion. They also found that low molar mass polymers, which may not be sufficiently large to provide complete stabilisation, were preferentially adsorbed. As most commercial resins contain a range of molar masses, some flocculation will always be obtained in practice, because of this preferential adsorption.

In addition to the thickness of the adsorbed layer, the concentration is also important. Factors which can affect concentration are chain length, and also the size, number and position of side chains. The solvent must be a good solvent for the polymer so that polymer–solvent contacts are more favourable energetically than polymer–polymer contacts.

Summarising, steric stabilisation will depend upon the polar groupings of the pigment surface and the polymer molecules. The polar groupings on the pigment surface are provided by the coatings. Some points of attraction must be present but not so many that a compact layer will be formed. Silica will provide acidic sites and alumina will act as a base.

The solvent must be sufficiently powerful to provide adequate solvation of the polymer. If polar solvents are present they must not be preferentially adsorbed on the pigment surface, although this will assist in the initial wetting prior to disruption. It is doubtful whether complete absence of flocculation is ever attained; electron micrographs inevitably show some flocculation.

Whilst freedom from flocculation will ensure complete dispersion and hence maximum opacity, in some instances a slight amount may be beneficial. It generally imparts a degree of thixotropy which may be useful in controlling flow. This thixotropy, together with the decrease in the effective specific gravity of the particles—the floccules comprise both pigment and resin solution—will prevent hard settlement. However, it should be rigidly controlled. In addition to reducing opacity, flocculation will also result in poorer gloss, poorer durability and also a yellower tone, as less blue light will be scattered.[52]

6.2.1. Measurement of flocculation

Balfour[52] and Franklin[53] describe a method, further developed by Rutherford and Simpson,[54] for giving a quantitative measurement of flocculation of titanium dioxide pigments. As previously mentioned when discussing opacity, the wavelength of light preferentially scattered is dependent upon the particle size. As floccules will act as a single scattering centre they will preferentially scatter radiation of longer wavelength. The greater the amount of flocculation, the greater the scattering of longer wavelengths. Thus, measuring the amount of infra-red radiation (2500 nm) scattered will give an indication of the degree of flocculation.

A method based on proton straggling has been described by Doroszkowski and Armitage.[55] This enables the size of the floccules in both the dry film and wet paint to be measured. Accordingly the elapsed time at which flocculation occurs can be determined.

6.2.2. Practical considerations[56]

With most millbases, wetting and disruption is favoured by a high solvent/low resin content. Pigment:binder ratios of 10:1 to 15:1 are not uncommon, with resin solution concentrations of 20 to 40%. As the oil absorption of titanium dioxide pigments is of the order of 20, there will be insufficient resin to provide a satisfactory barrier, and flocculation, and probably re-agglomeration, will occur on removal of the shear forces. Addition of further resin solution to the millbase, to give a pigment:binder ratio of 5:1, or preferably 4:1, while shear is still being applied, will result in a stable base. Great care is required at this initial 'let down', e.g. the concentration of the added resin solution should not be too dissimilar from that in the millbase to prevent re-agglomeration, often referred to as 'colloidal shock'.

Even when a stable dispersion has been obtained, flocculation can occur on addition of further components. The reason for this lies in the dynamic nature of the adsorption, which can result in the replacement of the stabilising polymer molecules by less efficient ones. Typical examples are:

(1) addition of thixotropic resins to give a structured paint,[57]
(2) addition of amino resins, some having no effect but others causing varying degrees of flocculation, which increases with amino resin content;

(3)   use of additives, particularly anti-settling agents.[58]

Care is also required that the nature of the adsorbed polymer should not be subsequently changed. This is particularly relevant with water-soluble resins, which may be supplied in acidic or basic form and subsequently made water-soluble by neutralisation. For dispersion stabilisation, it is preferable that the neutralised from be present in the millbase.

With electrodeposition paints the unneutralised form is often used to facilitate control of bath pH, but the possible adverse effect on dispersion stability on subsequent neutralisation should also be taken into consideration.

7. INTERACTIONS DUE TO PIGMENTS

The acid–base characteristics of the silica–alumina coatings in the presence or absence of zinc can cause interactions. Whilst some can be readily explained, other interactions are apparently contradictory and there are no obvious answers.

7.1. Effect on Cure

7.1.1. Resin systems involving catalysed crosslinking with amino resins
With resin mixtures which require only heat to cause satisfactory crosslinking, the degree of crosslinking is not affected by the grade of pigment used. This is not so when catalysts are used to effect cure at lower temperatures.

Two such systems in common use are the following:

(a)   High-solids resins, introduced over the past few years in order to reduce atmospheric pollution from solvent emission. The single-pack systems comprise polyesters, alkyds and acrylics crosslinked with hexamethoxymethyl melamine (HMMM) resins. Because of the low reactivity of these amino resins an acid catalyst is necessary to give the desired cure at an acceptable temperature.

(b)   Cold-cured alkyd/urea–formaldehyde (UF) finishes, used mainly on wood or other surfaces where heating to effect cure is not possible. They consist of highly reactive UF resins, plasticised generally by alkyds, crosslinking at room temperature is brought about by addition of acid catalysts.

With both resin systems the cure is affected by the grade of pigment used, but the pigment parameters having the greatest effect are not consistent between the two systems. With high-solids resins [59,60] with pigments free of zinc, there is a definite correlation between reduction in cure and alumina content. With zinc-rutilised grades there is no similar correlation with alumina but higher silica levels are detrimental. Zinc-rutilised grades often give slightly poorer cure than zinc-free pigments with similar coatings.

With cold cured alkyd/UF coatings,[60] zinc is the dominant factor, its presence resulting in poorer cure. With zinc-free grades the presence of alumina certainly does not reduce the cure, there being a fairly high optimum value (ca 4%) below and above which the cure is reduced. As with high-solids systems, higher silica levels give poorer cure with zinc-rutilised grades.

The reduction in cure with zinc-rutilised grades can be explained by partial neutralisation of the acid catalyst by zinc. The effect of basic alumina on the cure of acid-catalysed high-solids resins can be explained similarly. Further evidence[61] for such a pigment coating–catalyst interaction is that with a range of pigments of varying alumina/silica content, the order of hardness is reversed, depending upon whether the crosslinking is acid-catalysed by p-toluenesulphonic acid or base-catalysed by tetrabutylammonium iodide. Pigments with predominantly alumina coating (basic sites) give softer films with the acid-catalysed resins, whilst pigments containing silica (acidic sites) give softer films with base-catalysed systems. Whilst the effect of zinc and alumina can be explained, remaining anomalies are:

(a) Alumina is detrimental only with high solids, and then only with zinc-free grades.

(b) With zinc-rutilised grades, silica, and not alumina, is detrimental.

(c) Although zinc reduces the cure with both systems, the effect is greater with the cold-cured alkyd/UF resin than with the high-solids system.

Possible explanations are listed below.

(a) Although in each case the reaction is between a hydroxy-containing resin and an amino resin, the HMMM resin is of low reactivity and requires both heat and a catalyst to give satisfactory cure, whereas the UF resin is highly reactive and only a catalyst is necessary. Accordingly the mechanisms of crosslinking are probably different.

(b)   The deposition of silica and alumina may be different, depending upon whether zinc is present or not, more alumina being present on the surface when zinc is absent and therefore more accessible to react with the acid catalyst.

(c)   Polymer adsorption on the pigment surfaces is different in the two resin systems.

In addition to providing steric repulsion the adsorbed polymers may also act as a barrier between the pigment surface and the catalyst. In high-solids resins the polymer chains are of low molar mass, whilst the alkyd resins, used in conjunction with the reactive UF resins, are of medium to high molar mass, and when adsorbed the latter will be much more extended from the pigment surface, thereby providing a more effective barrier. This would certainly be the case with alumina-coated pigments (see Section 6.2).

(d)   The strength of the adsorption of the polymer acid groups on to the alumina sites may be lower with high-solids resins, and therefore more easily replaced by the acid catalyst.

(e)   The silica sites adsorb the reactive groups of the amino resin, preventing them from taking part in the crosslinking reaction.

Evidence for the importance of the adsorbed polymer configuration is provided by the frequently observed improvement in cure, in addition to better gloss and opacity, i.e. less flocculation, when the amount of butanol is increased in acid-catalysed alkyd/UF finishes pigmented with zinc-rutilised grades.

Probably there is no single reason for each observed anomaly but several contributory factors.

7.1.2. Powder coatings

The grade of pigment used can also affect the cure of powder coatings[60] based on polyesters crosslinked with either epoxy resins or triglycidyl isocyanate (TGIC), The cure of powder coatings consisting of epoxy resins crosslinked by amines or amides does not appear to be affected by the grade of pigment.

The dominant factor is the presence or absence of zinc but the effect is also very dependent upon the polyester resin used, zinc-rutilised grades giving poorer cure than zinc-free grades with some polyester resins but the order being reversed with others. Inclusion of colloidal zinc in the powder has a similar effect, but it is not as great as when the zinc is contained in the titanium dioxide pigment.

The silica–alumina coating also has a slight effect on cure but it is dependent upon the base pigment. Insufficient work has been reported to give a definite answer but the indications are that with zinc-rutilised grades, silica slightly improves the cure whilst alumina gives poorer cure. With zinc-free grades, silica is slightly detrimental but alumina has no effect. The reason for the definite effect of zinc, and the lesser effect of the coatings, on the cure of polyester powders is not clear. The most probable answer is the type of catalyst used during the manufacture of the resins.

7.1.3. Air-drying paints

Although the presence of zinc or the alumina/silica ratio does not affect the drying of paint containing metallic salts as driers, the grade of pigment may do so. For example, heavily coated grades may adsorb the driers due to the porous nature of the coating, typically 30 m^2/g, and when amines have been used to assist micronising they may form a complex with the cobalt drier. In each case, an increase of drier content will overcome the problems, although the colour may be adversely affected.

7.2. Effect on Flotation

Whilst dispersion stabilisation is not dependent on the charge on the pigment surface, the signs of the charges on white and coloured pigments are relevant with regard to flotation. Co-flocculation and hence flotation are much more likely if the charges are of opposite sign.[50] Whilst increasing the alumina content of the coating on titanium dioxide pigments will make the charge more positive (or less negative) the charge is also dependent upon the resin solution. Changing the resin solution may result in the charge on the titanium dioxide changing sign so that a pigment blend which is free from flotation in one resin solution may not be so if resin and/or solvent are changed.

7.3. Effect on Viscosity Stability

The presence of zinc can cause viscosity instability with some acidic resins. The problem is not very great with the majority of resins used in paints and inks but zinc-rutilised grades should be avoided in systems such as acid-modified vinyl resins and wash primers containing phosphoric acid.

Heavily coated grades, primarily intended for emulsion paints, can also give rise to viscosity instability in solvent-based systems. The instability appears to be a mixture of chemical reaction and physical adsorption producing a pigment/polymer/pigment network. With acid values below approximately 15 no problems arise, whereas at values above 25, heavily coated grades frequently give significant viscosity increases. Between acid values of 15 and 25 the results are very dependent upon the resin. If the paints are stored at elevated temperatures (40–60°C), as are frequently used for accelerated storage tests, the viscosity increase is often less than at room temperature, probably because of greater mobility of polymer molecules adsorbing and desorbing from the pigment surface. If the instability was caused by a chemical reaction, then the viscosity increase would be greater at higher temperatures.

ACKNOWLEDGEMENT

The editors of this volume thank Dr L. A. Simpson, Assistant Technical Service Manager of Tioxide Group plc, for help in preparing the manuscript of the late Mr T. Entwistle, for publication.

REFERENCES

The references are illustrative and not intended as comprehensive.

1. CALLOW, D. M., *Industrial Minerals*, 1985, **209**, 59.
2. WHITEHEAD, J., in *Kirk Othmer: Encyclopedia of Chemical Technology*, 3rd edn, Vol. 23, 1983, Wiley-Interscience, New York, 131.
3. BLECHTA, V. and LAVICKA, M., *J. Oil Colour Chemists Assoc.*, 1966, **49**, 195.
4. BLECHTA, V. and LAVICKA, M., *J. Oil Colour Chemists Assoc.*, 1967, **50**, 495.
5. HUGHES, W., *J. Oil Colour Chemists Assoc.*, 1952, **35**, 535.
6. EVANS, A. W. and MURLEY, R. D., *VI Fatipec Congress, Wiesbaden, 1962*, p. 125; *Paint Technol.*, 1962, **26**(10), 16.
7. HOWARD, P. B., British Patent 1479988.
8. HOWARD, P. B., British Patent 1479989.
9. HOWARD, P. B., British Patent 1481151.
10. WISEMAN, T. J. and HOWARD, P. B., British Patent 1365411.
11. WISEMAN, T. J. and HOWARD, P. B., British Patent 1365412.
12. ALLEN, A., US Patent 3897261.
13. DECOLIBUS, R. L., US Patent 3928057.
14. SCHMIDT, P. G., US Patent 3941603.
15. WISEMAN, T. J., in *Characterisation of Powder Surfaces*, ed. G. D. Parfitt and K. S. W. Sing, 1976, Academic Press, London, Chapter 4.
16. PARFITT, G. D., *Croat. Chem. Acta*, 1980, **52**, 333.
17. DAY, R. E., *Prog. Org. Coatings*, 1973, **2**, 269.
18. EGERTON, T. A., PARFITT, G. D., YOONOK KANG and WIGHTMAN, J. P., *Colloids & Surfaces*, 1983, **7**, 311.

222   T. ENTWISTLE

19. BLAKEY, R. R., *J. Oil Colour Chemists Assoc.*, 1983, **66**, 297.
20. RAYLEIGH, Lord, *Phil. Mag.*, 1871, **41**, 447.
21. MIE, G., *Phys. Lpz.*, 1908, **25**, 377.
22. KUBELKA, P. and MUNK, F., *Z. Techn. Phys.*, 1931, **12**, 593.
23. KUBELKA, P., *J. Opt. Soc. Amer.*, 1948, **38**, 448.
24. VAN DE HULST, H. C., *Light Scattering by Small Particles*, 1937, John Wiley and Sons, New York.
25. JUDD, D. B. and WYSZEIKI, G., *Colour in Business, Science and Industry*, 2nd edn, 1963, John Wiley and Sons, New York.
26. KERKER, M., *The Scattering of Light and Other Electromagnetic Radiations*, 1969, Academic Press, New York.
27. MITTON, P. B., in *Pigment Handbook*, Vol. III, ed T. C. Patton, 1973, Wiley Interscience, New York.
28. SIMPSON, L. A., *Measuring Opacity, TiINFO*, Section 3, Tioxide Group PLC.
29. ORCHARD, S. E., *J. Oil Colour Chemists Assoc.* 1968, **51**, 44.
30. ALLEN, E., *J. Paint Technol.*, 1973, **45**(584), 65.
31. ROSS, W. D., *J. Paint Technol.*, 1971, **43**(563), 50.
32. TUNSTALL, D. F. and DOWLING, D. G., *J. Oil Colour Chemists Assoc.*, 1971, **54**, 1007.
33. TUNSTALL, D. F. and HIRD, M. J., *J. Paint Technol.*, 1974, **46**(588), 33.
34. TUNSTALL, D. F., British Patent 2046898.
35. SIMPSON, L. A., *Prog. Org. Coatings*, 1978, **6**, 1; *Measuring Gloss, TiNFO*, Section 3, Tioxide Group PLC.
36. PEACOCK, J., *XI FATIPEC Congress, Florence*, 1972, p. 193.
37. HUGHES, W., *X FATIPEC Congress, Montreaux, 1970*, p. 67.
38. EGERTON, T. A. and KING, C. J., *J. Oil Colour Chemists Assoc.*, 1979, **62**, 386.
39. COLLING, J. H. and DUNDERDALE, J., *Proc. 6th Internat. Conf. in Org. Coatings, Science and Tech., Athens, 1980*, p. 239.
40. VOLZ, H. G., KAEMPF, G., FITZKY, H. G. and KLAEREN, A., *Amer. Chem. Soc. Symp. Ser.*, ed. S. P. Pappos and F. H. Winslow, 1981, **151**, 163.
41. SIMPSON, L. A., *Austral. OCCA Proc. News*, 1983, **20**(5), 6.
42. HARVEY, P. R., RUDHAM, R. and WARD, S., *J. Chem. Soc. Faraday Trans.*, 1983, **79**, 1381.
43. MURLEY, R. D., *J. Oil Colour Chemists Assoc.*, 1962, **45**, 6.
44. GERSTNER, W., *J. Oil Colour Chemists Assoc.*, 1966, **49**, 954.
45. PARFITT, G. D., *XIV FATIPEC Congress, Budapest, 1978*, p. 107.
46. CROWL, V. T., *J. Oil Colour Chemists Assoc.*, 1963, **46**, 169.
47. CROWL, V. T. and MALATI, M. A., *Disc. Faraday Soc.*, 1966, **42**, 301.
48. FRANKLIN, M. J. B., GOLDSBROUGH, K., PARFITT, G. D., and PEACOCK, J., *J. Paint Technol.*, 1970, **42**(551), 740.
49. GOLDSBROUGH, K. and PEACOCK, *J. Oil Colour Chemists Assoc.*, 1971, **54**, 506.
50. FRANKLIN, M. J. B., *J. Oil Colour Chemists Assoc.*, 1968, **51**, 499.
51. SMITH, A. L., in *Dispersion of Powders in Liquids*, ed. G. D. Parfitt, 3rd edn, 1981, Applied Science, London.
52. BALFOUR, J. G., *J. Oil Colour Chemists Assoc.*, 1977, **60**, 365.
53. BALFOUR, J. G. and FRANKLIN, M. J., *J. Oil Colour Chemists Assoc.*, 1975, **58**, 331.
54. RUTHERFORD, D. J. and SIMPSON, L. A., *J. Coatings Technol.*, 1985, **57**(724), 75.

55. DOROSZKOWSKI, A. and ARMITAGE, B. H., *XVII FATIPEC Congress, Lugano, 1984*, Vol. II, p. 311.
56. RACKHAM, J. R., *Pig. Res. Tech.*, 1976, No. 4, 8; No. 5, 10; No. 6, 6; No. 7, 11; No. 9, 3; also *TiINFO*, Section 3, Tioxide Group PLC.
57. HALL, J. E., *J. Coatings Technol.*, 1983, **55**(705), 41.
58. BLAKEY, R. R., *J. Paint Technol.*, 1971, **43**(559), 65.
59. BERTRANS, C. and GOSSELIN, E., paper presented at *6th Internat. Conf. in Organic Coatings, Science and Tech.*, Athens, July 1978.
60. ENTWISTLE, T. and GILL, S. J., *J. Oil Colour Chemists Assoc.*, 1986, **69**, 25.
61. BERTRAND, C., *XV FATIPEC Congress, Amsterdam, 1980*, Vol. II, p. 307; also Tioxide Technical Report D9105GC.

Index